Modern Applied Principles of
Thermal Power Plants

Modern Applied Principles of Thermal Power Plants

Edited by **Matt Danson**

LANRYE
INTERNATIONAL

New Jersey

Published by Clanrye International,
55 Van Reypen Street,
Jersey City, NJ 07306, USA
www.clanryeinternational.com

Modern Applied Principles of Thermal Power Plants
Edited by Matt Danson

International Standard Book Number: 978-1-63240-356-8 (Hardback)

Contents

Preface

A facility for the production of electrical energy from thermal energy released by combustion of a fuel or consumption of a fissionable material is known as a Thermal Power Plant. Thermal power plants are significant process industries for engineering specialists. The power sector has been facing several crucial issues over the past few years. The primary challenge is to meet the increasing power demand in a sustainable and efficient manner. Practicing power plant engineers not only look after the maintenance and operations of the plant, but also look after a variety of activities like research and development, starting from power generation to the environmental facets of the power plants. This book discusses topics like evaluation of plant performance, combustion, energy efficiency, catalytic reduction of dissolved oxygen, environmental facets of combustion residues and renewable power generation. It also elucidates issues related to both coal fired and steam power plants. It will be helpful for undergraduate and research oriented students, and for engineers working in power plants.

All of the data presented henceforth, was collaborated in the wake of recent advancements in the field. The aim of this book is to present the diversified developments from across the globe in a comprehensible manner. The opinions expressed in each chapter belong solely to the contributing authors. Their interpretations of the topics are the integral part of this book, which I have carefully compiled for a better understanding of the readers.

At the end, I would like to thank all those who dedicated their time and efforts for the successful completion of this book. I also wish to convey my gratitude towards my friends and family who supported me at every step.

Editor

Energy Efficiency and Plant Performance

Exergy Analysis and Efficiency Improvement of a Coal Fired Thermal Power Plant in Queensland

R. Mahamud, M.M.K. Khan, M.G. Rasul and
M.G. Leinster

Additional information is available at the end of the chapter

1. Introduction

Energy security and CO_2 emission reduction are two major concerns of today's world. Improving efficiency of the energy systems is an essential option for the security of future energy and the reduction of CO_2 emissions. With the growing prosperity of the civilization, our consumption of energy is growing very rapidly. Fossil fuels remain the world's dominant primary energy supply, with its use as a versatile primary source of energy in power generation, transport and industry. However, we have finite sources of these non renewable fossil fuels and we are consuming them at a rate that cannot be sustained, leading to the risk on energy security in the future. The Intergovernmental Panel on Climate Change (IPCC), in its Fourth Assessment Report [3], identified carbon dioxide (CO_2) emissions from burning of fossil fuels as the primary contributor to climate change. Therefore, the prudent use of energy is imperative and the importance of improving energy efficiency is now very well realized by all.

Improvement of energy efficiency of power generation plants leads to lower cost of electricity, and thus is an attractive option. According to IEA (IEA 2008a), the average efficiency of coal-fired power generators of the Organisation for Economic Co-operation and Development (OECD) member[1] countries over the 2001 to 2005 period in public sector is 37%. According to this report, the highest average efficiency of coal-fired power plants is observed in Denmark which is 43% and in United States 36%. The average energy efficiency of Australian coal-fired power plants is one of the lowest among the OECD countries which is 33%. Therefore, improving energy efficiency of coal-fired power plant in Australia is very important.

---

1 OECD members are economically developed countries.

The energy conversion in a coal-fired power plant is dominantly a thermodynamic process. The improvement of energy efficiency in a thermodynamic process generally depends on energy analysis of the process which identifies measures required to be addressed. The conventional method of energy analysis is based on first law of thermodynamics which focuses on conservation of energy. The limitation with this analysis is that it does not take into account properties of the system environment, or degradation of the energy quality through dissipative processes. In other words, it does not characterize the irreversibility of the system. Moreover, the first law analysis often casts misleading impressions about the performance of an energy conversion device [4-6]. Achieving higher efficiency, therefore, warrants a higher order analysis based on the second law of thermodynamics as this enables us to identify the major sources of loss, and shows avenues for performance improvement [7]. Exergy analysis characterizes the work potential of a system with reference to the environment which can be defined as the maximum theoretical work that can be obtained from a system when its state is brought to the reference or "dead state" (standard atmospheric conditions). The main purpose of exergy analysis is to identify where exergy is destroyed. This destruction of exergy in a process is proportional to the entropy generation in it, which accounts for the inefficiencies due to irreversibility.

This research conducts exergy analysis in one unit of a coal-fired power plant in Central Queensland, Australia as a case study. The exergy analysis identifies where and how much exergy is destroyed in the system and its components. Based on the analysis, it assesses and discusses different options to improve the efficiency of the system.

2. Process description of a coal-fired power plant

A coal-fired power plant burns coal to produce electricity. In a typical coal-fired plant, there are pulverisers to mill the coal to a fine powder for burning in a combustion chamber of the boiler. The heat produced from the burning of the coal generates steam at high temperature and pressure. The high-pressure steam from the boiler impinges on a number of sets of blades in the turbine. This produces mechanical shaft rotation resulting in electricity generation in the alternator based of Faraday's principle of electromagnetic induction. The exhaust steam from the turbine is then condensed and pumped back into the boiler to repeat the cycle. This description is very basic, and in practice, the cycle is much more complex and incorporates many refinements.

A typical coal plant schematic is presented in Figure 1. It shows that the turbine of the power plant has three stages: high-pressure, intermediate-pressure and low-pressure stages. The exhaust steam from the high-pressure turbine is reheated in the boiler and fed to the inter-mediate-pressure turbine. This increases the temperature of the steam fed to the intermediate-pressure turbine and increases the power output of the subsequent stages of the turbine. Steam from different stages of the turbine is extracted and used for boiler feed water heating. This is regenerative feed water heating, typically known as regeneration. The improvement of the thermal performance of the power generation cycle with reheat and regeneration is a trade-off between work output and heat addition [8] and it can be evaluated through the efficiency of the power generation cycle.

In a typical pulverised coal power plant, there are three main functional blocks as shown in Figure 1. They are (1) the boiler; (2) the turbo-generator and (3) the flue gas clean up. The boiler burns coal to generate steam. The combustion chamber of the boiler is connected with the coal pulverisers and air supply. The water pre-heater (also known as the economiser), the super heater and the reheater are all included in this block. The steam produced in the boiler is used in the turbine as shown in Figure 1. The generator is coupled with the turbine where mechanical shaft rotation of the turbine is converted into electrical power and supplied to the power distribution grid through a transformer. The purpose of the transformer is to step up the voltage of the generated power to a level suitable for long distance transmission. The steam leaving the turbine is condensed in the condenser as shown in the Figure 1 using cooling water which discharges low temperature heat to the environment. The condensate produced is pumped back to the boiler after heating through the feed water heaters. The feed water heaters use regenerative steam extracted from the turbine.

The burning of coal in the boiler of a power plant produces flue gas. The main constituents the of flue gas are nitrogen (N_2), carbon dioxide (CO_2) and water (H_2O). It carries particulate matter (PM) and other pollutants. There are traces of some oxides such as oxides of sulphur (SOx) and oxides of nitrogen (NOx) depending on the combustion technology and fuel used. The flue gas clean-up block comprises all the equipment needed for treating the flue gas. The power plant shown in Figure 1 includes a DeNOx plant for NOx removal, followed by electrostatic precipitation (ESP) to remove particulate matter (PM), and wet flue gas desulfurisation (FGD) to remove SOx from the flue gas. An air-preheating unit is situated between the DeNOx and the electrostatic precipitator (ESP). There is a significant amount of heat energy leaving through the flue gas, some of which is recovered by using the air preheater. This improves the thermal performance of the process.

The properties of the coal used in the boiler and the environmental legislation and/or environmental management policy of a plant are two major factors that determine the nature of the flue gas treatment process. In some countries, due to stringent environmental regulation, coal-fired power plants need to install denitrification plants (DeNOx) for nitrogen oxide (NOx) and flue gas desulphurisation plants (FGD) for sulphur oxide (SOx) removal [9, 10]. In Australia, the coal used has a very low sulphur content and therefore, the concentration of SOx from the burning of coal in Australia is relatively low. Dave et al. [11] report an absence of stringent regulatory requirements for limiting NOx or SOx in flue gas streams in Australia. Therefore, Australian coal plants in the past have not been required to have deNOx or deSOx equipment to clean up flue gas.

In this research, a pulverized coal-fired power plant in Central Queensland, Australia has been considered as a case study. One of the units of the said plant was used to develop a process model and to perform energy analysis. This unit has Maximum Continuous Rating (MCR) of 280 MW. It spends less that 5% of its operating time at loads greater than 260 MW. Operation of the unit is mostly in the range of 100 to 180 MW range. The unit plant is a sub-critical power

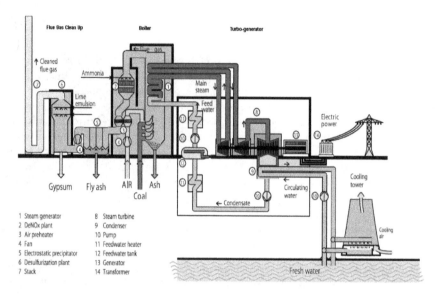

Flue Gas Clean Up   Boiler   Turbo-generator

1	Steam generator	8	Steam turbine
2	DeNOx plant	9	Condenser
3	Air preheater	10	Pump
4	Fan	11	Feedwater heater
5	Electrostatic precipitator	12	Feedwater tank
6	Desulfurization plant	13	Generator
7	Stack	14	Transformer

Figure 1. A typical Coal Power Plant

plant having steam outlet pressure of 16.2 MPa. The unit/plant uses thermal coal supplied from the nearby Bowen basin.

3. Process modeling and simulation

Mathematical models are effective tools for analysing systems or processes. They can be used to develop a new system or to evaluate the performance of an existing one. Mathematical modelling is widely applied to the solution of engineering problems. Modelling usually describes a system using a set of variables and equations and sets up relationships among the variables. Mathematical models are found to be very useful in solving problems related to process energy efficiency and can be utilised for both static and dynamic systems. SysCAD [1], a process modelling software package, has been found to be very effective for analysing plants for efficiency improvement. It has been used in Australia by a number of process industries, consulting companies and universities as a tool for simulating plant. Therefore, SysCAD has been employed in this study for modelling and simulating the said coal-fired power plant.

3.1. Modelling in SysCAD

SysCAD can work in both static and dynamic modes. In static mode, it can perform process balances known as ProBal. SysCAD process modelling in ProBal is illustrated in Figure 2. It shows the overall approach of developing a process model in SysCAD. A process model is generally treated as a project in SysCAD. A project may have one process flow sheet or a number of flow sheets. In a project, the flowsheets can interchange data and can be interlinked.

SysCAD uses typical graphical techniques to construct a process flow diagram (PFD). Before constructing a PFD in SysCAD flowsheet, the scope of the project and data required to perform modelling needs to be defined as shown in the Figure 2. In SysCAD there are many built-in process components known as unit models. The components in a process are represented by the unit models. For modelling purposes, these unit models need to be configured based on the system requirements and performance data of the individual components used in the process as shown in Figure 2. All the unit models need to be connected appropriately to construct the PFD in a flowsheet. There are chemical species defined in SysCAD, which are used to calculate physical and chemical properties. The user can also define process components and chemical species as required if they are not available in the SysCAD component library or species database. Chemical reactions are important to perform some process modelling. SysCAD has built in features to define and simulate chemical reactions. Modelling with chemical reactions requires defining all chemical reactions in a reaction editor. The extent of a reaction based on a certain reactant can be provided as a fraction. If multiple reactions are required to define a model, the sequence of each reaction can be provided. SysCAD uses a user-defined sequence during simulation.

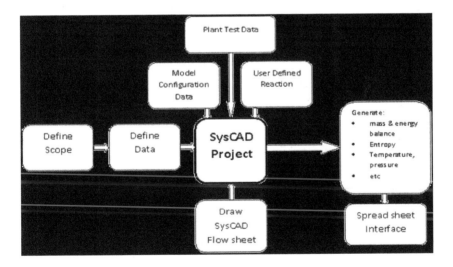

Figure 2. SysCAD Modeling [2]

As shown in Figure 2, SysCAD ProBal function provides a mass and energy balance of a process and its components. The mass balance incorporates all input and output streams together with any mass additions or sinks in unit operations. This balance considers changes due to reactions or phase changes. An energy balance, on the other hand, looks at the input and output streams as well as all sources of heat transfer simply via total enthalpy. The concept is that each stream has a sum of total enthalpy and if a unit operation is balanced the sum of total enthalpy at T_{in} of all respective streams equals the sum of total enthalpy at T_{out} of all respective streams.

In addition to energy and mass balances, SysCAD offers a wide range of thermo-physical data as shown in Figure 2, which is important for analysing the process and its components (e.g., temperature, pressure, entropy). SysCAD has some common process control models such as PID controller, actuator, transmitter and general controller. The general controller can be defined and used through a built-in programming language called Programmable Module (PGM). It has extended the functionality of the process modelling in SysCAD. Process data can be used in PGM to control a process and to perform calculations required for useful analysis. In addition, data from a process balance can be exported to MS Excel to perform further analysis.

In an efficiency improvement study, process energy analysis is very important for identifying where energy is lost and how effectively the loss can be minimised or recovered in the process. The results of a process modelling and simulation in SysCAD can be further analysed to observe, identify and assess energy lost in a process.

3.2. Brief description of case study

In this study, a power plant model developed based on a unit of a power plant in Central Queensland Australia. The power plant uses pulverised coal supplied through pulverisers and burnt in a boiler. The boiler of the plant is of the radiant tube type. It has natural circulation design with a low and high temperature economiser, a three-stage superheater and a two-stage single reheater. The boiler has a maximum steam outlet pressure of 16.2 MPa and a temperature of 541°C and feed water maximum temperature of 252° C.

The unit plant has a turbine to convert to convert thermal energy of steam into mechanical shaft rotation. The turbine has three pressure stages – high pressure, intermediate pressure and low pressure. In all three stages, there are stream extractors to facilitate regenerative heating of feed water heater. Three low-pressure heaters (LPH), one deaerator and two high-pressure heaters (HPH) use bled steam for regenerative feed heating. A condenser is used in the power plant to condense low pressure steam into water. The condenser is water-cooled type and it has been built for seawater operation. There is a minor loss of water in the plant process. Therefore, makeup water for the boiler feed is added into the condenser hot-well after passing through a deaerating system. It has been found that the requirement for makeup water in the boiler is very low compared to the total requirement. The condensate passes through a series of heat exchangers - LPHs, deaerator and HPHs which take heat from the regenerative bled stream as mentioned earlier.

The highest capacity of the power plant is 280 MW of electrical power. This is the maximum capacity rating (MCR) of the power plant. The capacities of all individual process components were configured with appropriate data to produce the rated power. A detailed description of the configuration of all individual process components is provided while describing the model flow sheets in the subsequent section.

3.3. Process model development

In this research, the power plant was represented by two separate flow sheets. They are a) Power Generation Model and b) Boiler Combustion Model. The detailed descriptions of these two model flow sheets are provided in the next sections.

3.3.1. Power generation model

The flow sheet for the power generation cycle is presented in Figure 3. It shows the steam cycle of the power plant. This cycle is known as the Rankine cycle [8] including reheating and regeneration.

The boiler in the power plant has feed water heater, superheater and reheater. The boiler feed water heater and super heater are included in the boiler model while the reheater is represented as a separate heater denoted as 'reheating' as shown in Figure 3.

The whole turbine is modelled using 7 unit turbine models as shown in the figure. This was done to simultaneously facilitate the use of the inbuilt SysCAD turbine model and steam extraction. For example, due to two steam extractions from intermediate pressure stage of the turbine, it is represented by two unit models namely IP_TRB1 and IP_TRB2.

The steam leaving the low-pressure turbine was connected with a condenser, which is described using a shell and tube type heat exchanger in SysCAD. The condenser is supplied with cooling water to perform steam condensation. The pressure of the steam at this stage is very low. The bled steam extracted from the turbine is recycled in to the condenser. The makeup water required in the process is added after the condenser. The condensate pump is located after the condenser, and it boosts the pressure of the condensate high enough to prevent boiling in the low-pressure feedwater heaters. The condensate mixes with the makeup water before entering the condensate pump.

There are three low-pressure heaters connected with the extracted steams from different turbine stages as shown in Figure 3. The feed water is gradually heated, taking heat from steam with increasing temperature and pressure at each stage. In a low-pressure heater, heat exchange occurs in two stages. At first, the steam condenses to its saturation temperature at steam pressure and then occurs sensible heat exchange. All the three heaters were developed based on the same principle.

The main purpose of a deaerator is to remove dissolved gases including oxygen from the feed water. Some heat exchange occurs in the deaerator. In this model, the deaerator is treated as a heat exchanging device where steam and feed water exchange heat through direct contact. The tank model built in SysCAD was used to represent the deaerator. The feed water, after heating in the deaerator, is pumped through the feed water pump. The high pressure feed water is heated through two more high-pressure heaters. The steam for the high pressure heaters is extracted from the high pressure turbine exhaust, and from an inter-stage bleed on the intermediate pressure turbine. Each of the high-pressure heaters is developed using two SysCAD heat exchange models as described earlier for the low-pressure heaters.

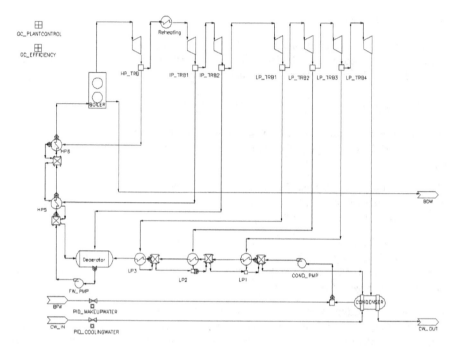

Figure 3. Power Plant Model

The operations of all the individual components used in the model flow sheet are described in detail in the subsequent discussion. The discussion includes data used to configure each component for the modelling.

3.3.1.1. Boiler and reheating

The boiler model in SysCAD calculated energy required by the boiler based on the boiler feed water, the required drum pressure and the superheated steam conditions. It is a very simple model, which does not take into account the type of fuel used or the type of economiser. It heats the high pressure feed water stream to saturation temperature and then to the super-heated condition as specified. A portion of saturated water is blown down from the boiler to discharge impurities and maintain water purity. In this research the operating conditions of the boiler were configured based on plant-supplied data presented below:

- Steam outlet pressure: 16 MPa

- Steam outlet temperature: 540 °C

- Boiler Efficiency: 90%

- Blow down: 0.5%

The high temperature steam from the boiler superheaters enters the high-pressure side of the turbine named HP_TRB. In this turbine stage, part of the steam energy is converted to mechanical shaft rotation and the pressure and temperature of the steam drops based on turbine configuration and supplied data discussed later. The steam leaving HP_TRB is taken to the reheating process where the steam is reheated to 540 °C. A portion of the steam is taken out to regeneration before it goes to reheating. The reheating model is a simple heater for sensible heating and no phase change occurs. The simple heater only calculates the heat required for reheating.

3.3.1.2. Steam turbine

The built-in steam turbine model in SysCAD transforms steam energy into electrical power. In a flow sheet, it needs to be connected with one single steam input and one single stream output. The inlet conditions of the steam, such as temperature, pressure, mass flow and quality of steam need to be defined. Using steam inlet data and specified turbine efficiency, SysCAD calculates turbine output power and the condition of the outlet stream. The simplified energy balance calculation against a turbine is provided in the following equation:

$$W = H_{in} - H_{out} \qquad\qquad (1)$$

where, W is work output of the turbine, and Hin and Hout are enthalpy of stream into and out of the turbine

The high-pressure turbine was modelled using a turbine unit model named HP_TRB. The steam exiting HP_TRB was connected with the reheating system. The reheater heats up the steam to a high temperature to improve the quality of steam. A portion of the steam is extracted before the reheater and fed to a high-pressure heater named HP6. In Figure 3, the reheated steam enters the intermediate-pressure turbine. This stage was modelled using two turbine unit models. The two models were named as IP_TRB 1 and IP_TRB 2. Bled steam from both models is taken out – from the IP_TRB 1 to the high pressure heater, HP5 and from IP_TRB 2 to the deaerator. The steam from IP_TRB2 is connected with the low-pressure turbine. The low-pressure turbine is defined using four interconnected models named LP_TRB 1, LP_TRB 2, LP_TRB 3 and LP_TRB 4. There are bled steam flows from LP_TRB 1, LP_TRB 2 and LP_TRB 3. The bled steam flows are connected with three low-pressure heaters namely, LP3, LP2 and LP1 consecutively as shown in Figure 3.

It should be noted here that the SysCAD turbine model ignored changes of potential and kinetic energy since the changes are negligible. The turbine efficiency, mechanical efficiency, outlet pressure and steam bleed of the turbine in different stages are configured with data supplied by the plant and provided in Table 1.

	HP Turbine Stage(s)	IP Turbine Stage(s)		LP Turbine Stage(s)			
Turbine efficiency (%)	85						
Mechanical Efficiency (%)	98						
Outlet Pressure (kPa)	4000	1450	800	230	90	20	10
Bleed (%)	10	3	7	5	5	2	

Table 1. Input Data for Turbine Configuration

3.3.1.3. Condenser

The condenser was represented by a shell and tube heat exchanger, and it transfers energy from one stream to another. The primary use of this model is to transfer latent heat by steam condensation. The model performed the following calculations as defined in SysCAD.

For the heat exchanger:

$$Q = UA\Delta T_{LM} \tag{2}$$

where

Q - Rate of Heat Transfer

U - Overall coefficient of Heat Transfer

A - Area available for Heat Transfer

ΔTLM- Log Mean Temperature Difference (LMTD) calculated as

$$\Delta T_{LM} = \frac{\Delta T_2 - \Delta T_1}{\ln (\Delta T_2 / \Delta T_1)}$$

For Counter Current Flow $\Delta T_2 = T_{H_{in}} - T_{C_{out}}$ and $\Delta T_1 = T_{H_{out}} - T_{C_{in}}$

Where

TH_{in} – temperature of hot stream in

TH_{out} – temperature of hot stream out

TC_{in} – temperature of cold stream in

TC_{out} – temperature of cold stream out

This has been described for a counter flow heat exchanger in Figure 4.

For the heat transfer to the individual stream

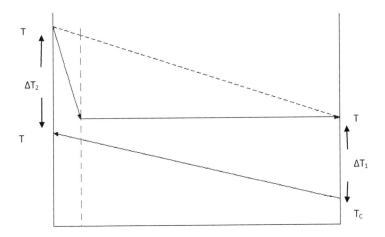

Figure 4. Log mean temperature difference [1]

$$Q = m\left(h_{in} - h_{out}\right) \tag{3}$$

where

Q - Rate of Heat Transfer

m - Mass flow of the stream

h_{in} – Specific enthalpy of entering stream

h_{out} – Specific enthalpy of leaving stream

In this model, the vapour entering the condenser first comes to the saturation temperature and is then condensed. No further cooling of the liquid occurs. The area of the heat exchanger and the cooling water required to condense the whole of the steam flow to the condenser are specified.

3.3.1.4. Low pressure and high pressure heaters

The low pressure and high-pressure heaters were developed primarily using the shell and tube heat exchanger model as described earlier. Each of the heaters uses two models to achieve its desired functionality. The first one was used to condense steam into saturated water and the second exchanger was used to cool the saturated water to a temperature below the saturation temperature through sensible cooling. In SysCAD, a heat exchanger model has only one desired functionality. If one heat exchange model performs condensation of steam from some temperature above the saturation temperature, it cannot perform any further cooling to the stream. Therefore, the second heat exchanger was used to achieve the desired functionality as required by the process. All the low-pressure heaters named LP1, LP2 and LP3 and the high-pressure heaters HP5 and HP6 were developed based on the same principle.

3.3.1.5. Deaerator

The fundamental purpose of a deaerator in power generation is to remove oxygen and dissolved gases from boiler feed water. This helps prevent corrosion of metallic components from forming oxides or other chemical compounds. However, in the power generation model the deaerator was treated as a direct contact heat transfer component in order to describe it for the desired purpose of this study. In the deaerator, steam comes in direct contact with liquid water and therefore heat transfer occurs.

A tank model in SysCAD is a multipurpose model. There are sub-models available with a tank model such as reaction, environmental heat exchange, vapour liquid equilibrium, heat exchange, make-up, evaporation, and thermal split. It was used here for defining the deaerator. This tank model was configured to achieve vapour liquid equilibrium only. The other sub-models were not used. The size of the deaerator tank was kept at 10 m^3 and all the streams were brought to the lowest pressure through a built-in flashing mechanism.

3.3.1.6. Pumps

There are two pump models used in the power generation model. One is a condensate pump and the other is a feed water pump. The pump model boosts pressure of liquid to a specified pressure. In order to configure a pump in SysCAD it needs to be connected with an incoming stream and an output stream. The energy balance across a pump was achieved through using the following equation:

$$W = H_{in} - H_{out} \tag{4}$$

It was assumed that the process is adiabatic and there were negligible changes of potential and kinetic energy, and they were therefore ignored. The two pump models were configured with their required pressure boost data as provided in Table 2.

Pump	Pressure Boost in kPa
Feed Water Pump	21670
Condensate Pump	1200

Table 2. Pump Configuration Data

3.3.1.7. Pipe

The pipe model in SysCAD is used to transfer material between two units. There is a large amount of information on the stream displayed in the pipe model. It also allows some user-defined calculations using data found on the pipe. The pipe model can take pipe friction loss into consideration. However, in this project the loss in the pipe was considered to be insignificant and was therefore ignored. In the pipe model, at different points of the power generation model code was applied to perform an exergy flow calculation. This is called Model Procedure (MP) in SysCAD and the code used here is similar to the codes in PGM described earlier. The details of exergy calculation are discussed later in section 4.2.

3.3.1.8. Controls and calculation of power generation model

There are two general controllers and two PIDs used to control the model to perform its set objectives. The general controllers were named as GC_PLANTCONTROL and GC_EFFICIEN-CY. GC_PLANTCONTROL was used to set the model simulation at the plant's desired rated capacity. Codes were also used to calculate the net power output and the feed water requirements. PID_MAKEUPWATER works in conjunction with GC_PLANTCONTROL to achieve the net output set point through controlling the boiler feed water. GC_EFFICIENCY was used to calculate overall efficiency of the power plant by dividing net power output by fuel energy input rate. The net power output was calculated by deducting power used in the pumps from the generated power in the turbine. PID_COOLINGWATER was used to regulate supply of the required amount of cooling to the condenser.

3.3.2. Boiler combustion model

The boiler combustion model was developed to supply the desired amount of heat to the boiler. The model flow sheet is presented in Figure 5. As shown in the figure, the main components of this model are a (boiler) combustion, a water heater (economiser), a superheater, a reheater and an air preheater. The combustion model was developed using a tank model, three heaters by simple heaters and the air preheater by a heat exchanger model built in SysCAD.

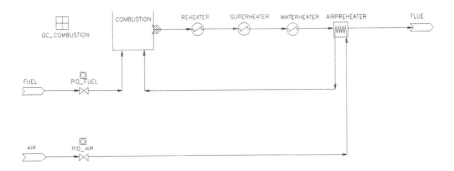

Figure 5. Boiler Combustion Model

3.3.2.1. (Boiler) combustion

The combustion model used a tank model built in SysCAD to perform the transformation of chemical energy to heat energy. A reaction sub-model was configured to perform that transformation. The chemical composition of the fuel was supplied by the plant and defined in the feed of the combustor named FUEL. The reactions and their extent were defined in a reaction editor of the tank model. The chemistry of combustion is complex and depends on many different factors. It was assumed that there is sufficient air supplied to complete the combustion of coal in air. Nevertheless, the stoichiometric chemical equations used here were placed in a logical order based on the chemical affinity of the components.

The power plant uses thermal coal supplied from nearby coal mines. The gross calorific value (GCV) of the coal at dry ash free (daf) conditions is 30.06 MJ/kg while at air dried (ad) conditions it is 20.80 MJ/kg. Data on the composition of the coal was supplied by the power plant and is presented in Table 3.

Proximate Analysis	Ultimate Analysis
Moisture 10.9%	Carbon 78.1%
Ash 18.9%	Hydrogen 3.9%
Volatile Materials 23.7%	Nitrogen 1.1%
Fixed carbon 46.5%	Sulphur 0.2%
	Oxygen 16.7%

Table 3. Coal Property Data

As mentioned earlier, chemical reactions were performed in reaction editor in SysCAD. In the reaction editor reaction, extent and sequence is provided. The combustion reaction for this modelling purpose is provided in Table 4.

Reaction	Extent	Sequence
$2 H_2(g) + 1 O_2(g) = 2 H_2O(g)$	Fraction $H_2(g) = 1$	1
$1 S(s) + 1 O_2(g) = 1 SO_2(g)$	Fraction $S(s) = 1$	2
$1 C(s) + 1 O_2(g) = 1 CO_2(g)$	Fraction $C(s) = 1$	3
$1 H_2O(l) = 1 H_2O(g)$	Fraction $H_2O(l) = 1$	4

Table 4. Combustion Reaction

Using the SysCAD property database the simulation calculates the heat of reaction (HOR) of each reaction in the combustor and then sums up all HORs to calculate overall HOR of full combustion. An environmental heat exchange is configured to allow for some heat lost to the environment from the combustor.

3.3.2.2. Water heater (economiser), superheater and reheater

These three components were modelled using the simple heater model built in SysCAD. The simple heater does not consider heating media or heater size. It only provides an estimation of heater duty required at stream outlet temperature or stream outlet temperature for specific heater duty. It can also be configured to specify heater duty irrespective of temperature. In this model, the heaters were configured to supply a specific amount of heat through heater duty calculated in the boiler and reheater in the power generation model by a so-called duty method in SysCAD. Only the duty of such heating was calculated for these heaters.

3.3.2.3. Air preheater

The air preheater in this flowsheet was built using the heat exchanger model described previously. The process is air/air heat transfer and the heat transfer coefficient used here is 150 w/m²K, which is lower than the value used for most of the liquid/liquid heat transfer.

3.3.2.4. Control and calculation in combustion model

In this model, three control elements were used. One was GC_COMBUSTION, a general controller and the other two were PIDs, one for fuel and the other for air. GC_COMBUSTION calculates the amount of fuel and excess air required to complete the combustion in the combustor. The two PIDs for fuel and air regulate the required fuel and excess air to achieve a set point. The set point for excess air was 10% as supplied by the plant. The fuel requirement of the combustion was set dependent on the energy requirements in boiler and reheating in power generation model. This has increased the functionality of the model to produce any set amount of power output. Similar to the power generation model, in pipes at different points of the boiler combustion model, code was applied to perform the exergy calculation using the Model Procedure (MP) of SysCAD.

4. Energy analysis and efficiency improvement

Energy analysis of a process is very important for identifying where energy is lost. It is performed through a process energy balance. This essentially considers all energy inputs in and outputs out of the system. When the system is balanced, the sum of all energy inputs equals the sum of all energy outputs. In a power generation plant, the objective is to convert the maximum possible energy input into useful work. According to the second law of thermodynamics, due to thermodynamic irreversibility not all energy input is converted in to useful work.

Traditionally, the energy analysis of a process is performed through energy balance based on the first law of thermodynamics. It focuses on the conservation of energy. The shortcoming of this analysis is that it does not take into account properties of the system environment, or degradation of energy quality through dissipative processes [12]. In other words, it does not take account of the irreversibility of the system. Moreover, the first law analysis often gives a misleading impression of the performance of an energy conversion device [4-6]. Getting an accurate estimate warrants a higher order analysis based on the second law of thermodynamics, as this enables us to identify the major sources of loss and shows avenues for performance improvement [7]. This essentially refers to exergy analysis that characterises the work potential of a system with reference to the environment.

4.1. Energy balance calculation

The energy balance was performed for the whole power generation process. As it was assumed that the energy lost in pipes is negligible, the loss of energy in process components represents the loss of energy of the whole process. The analysis was, therefore, performed to balance energy flow against all process components such as the boiler, turbine, and heat exchangers. This calculation provided information about where and how much energy is lost. The process model developed in SysCAD performs mass and energy balances considering all the input and output streams and heat and work into and out of each component. The equations used for these balances are provided in Equations 5, 6 and 7.

The mass balance for a unit process

$$\sum_{i=1}^{i} m_i = \sum_{o=1}^{o} m_o \tag{5}$$

where m is mass flow rate in Kg/s

The energy balance for a unit process

$$\sum_{i=1}^{i} E_i + \dot{Q} = \sum_{o=1}^{o} E_o + \dot{W} \tag{6}$$

where E is energy flow, $\dot{Q}$ heat flow and $\dot{W}$ is work flow in Kg/s

The energy flow of the stream was calculated as

$$E = \dot{m}h \tag{7}$$

where h is specific enthalpy in kJ/Kg and the potential and kinetic energy of the stream are ignored.

In order to obtain a balance of energy flow against different components of the power plant Figure 6 is used. The points in these figures were chosen very carefully so that they could describe the inflow and outflow of energy carried by streams to and from each component. The work inflows and outflows were observed from the SysCAD process energy balance and were used in balance calculations where they applied.

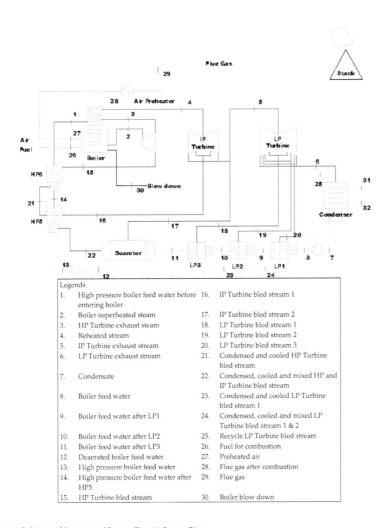

Figure 6. Points of Energy and Exergy Flow in Power Plant

The energy flows at the mentioned points were obtained directly from the SysCAD process balance at each mentioned point and exported to Microsoft Excel. It is important to note here that as potential and kinetic energy of the stream was ignored, SysCAD calculated the energy flow of the stream using Equation 7. The work in and out of the components, particularly in different stages of turbine and pumps, were found in the SysCAD process balance. Therefore, the balance of energy flow was calculated observing all energy into and out of each component in the form of either heat or work. The details of energy balance calculations were performed in Microsoft Excel across various process equipment using equations provided in Table 5.

Components	Energy Balance Equation
Boiler	$E_{26} + E_{27} + E_1 + E_3 = E_2 + E_4 + E_{28} + E_{30} + E_{I(Boiler)}$
HP Turbine	$E_2 = E_3 + E_{15} + W_{(HP\ Turbine)} + E_{I(HP\ Turbine)}$
IP Turbine	$E_4 = E_5 + E_{16} + E_{17} + W_{(IP\ Turbine)} + E_{I(IP\ Turbine)}$
LP Turbine	$E_5 = E_6 + E_{18} + E_{19} + E_{20} + W_{(IP\ Turbine)} + E_{I(LP\ Turbine)}$
Condenser	$E_{25} + E_6 = E_7 + E_{I(Condenser)}$
Condensate Pump	$E_7 + W_{(Condensate\ Pump)} = E_8 + E_{I(Condensate\ Pump)}$
LP1	$E_{24} + E_{20} + E_8 = E_9 + E_{25} + E_{I(LP1)}$
LP2	$E_{23} + E_{19} + E_9 = E_{10} + E_{24} + E_{I(LP2)}$
LP3	$E_{18} + E_{10} = E_{11} + E_{23} + E_{I(LP3)}$
Deaerator	$E_{22} + E_{17} + E_{11} = E_{12} + E_{I(Deaerator)}$
Feed Pump	$E_{12} + W_{(Feed\ Pump)} = E_{13} + E_{I(Feed\ Pump)}$
HP5	$E_{22} + E_{21} + E_{16} = E_{14} + E_{22} + E_{I(HP5)}$
HP6	$E_{14} + E_{15} = E_{21} + E_{I(HP6)}$

Table 5. Energy Balance in Power plant and Capture Process

Here, E represents energy flow in kW. The subscripts used in the energy balance represent point numbers in Figure 6. E with subscript l represents energy lost and the corresponding process is mentioned in the subscript within the bracket. W represents work and the corresponding process component is mentioned in the subscript within the bracket.

4.2. Exergy balance calculation

Exergy can be defined as 'work potential', meaning the maximum theoretical work that can be obtained from a system when its state is brought to the reference or "dead state' (under standard atmospheric conditions). The main purpose of exergy analysis is to identify where exergy is destroyed. This destruction of exergy in a process is proportional to the entropy generation in it, which accounts for the inefficiencies due to irreversibility. Exergy analysis helps in identifying the process of irreversibility leading to losses in useful work potential and thus pinpointing the areas where improvement can be sought.

Rosen [6] identified various exergy studies conducted by different researchers and found the significance of application for process energy analysis whether it is small or large but particularly more important for energy intensive ones e.g., power generation where large scale energy conversion happens. From this point of view, exergy analysis for power plants can be useful for identifying the areas where thermal efficiency can be improved. It does so by providing deep insights into the causes of irreversibility.

The exergy is thermodynamically synonymous to 'availability' of maximum theoretical work that can be done with reference to the environment. Som and Datta [13] define specific exergy, a (in kJ/Kg) in general terms as in Equation 8.

$$a = k + \varnothing + (u - u_r) + p_r(v - v_r) - T_r(s - s_r) + a_{ch}$$
$$\underbrace{\qquad\qquad\qquad\qquad}_{\text{Physical Exergy}} \quad \underbrace{\quad}_{\text{Chemical Exergy}}$$

(8)

where k (in kJ/kg) is the specific kinetic energy of the system and $\varnothing$ (in kJ/kg) is the potential energy per unit mass due to the presence of any conservative force field. T (in K), p (kN/m2), u (in kJ/kg), v (m3/Kg) and s (kJ/Kg-K) are the temperature, pressure, specific internal energy, specific volume and specific entropy, respectively, while a_{ch} represents the specific chemical exergy. The terms with the subscript r are the properties of the exergy reference environment. The generalised equation above can be simplified or specified based on a process. The kinetic and potential energies are small in relation to the other terms, and therefore they can be ignored.

The exergy flow was calculated at different points before and after the process components in different streams. The exergy of a stream was treated for both physical and chemical exergy. The physical exergy accounts for the maximum amount of reversible work that can be achieved when the stream of a substance is brought from its actual state to the environmental state.

According to Amrollahi et.al. [14] this can be evaluated as

$$a = (h - h_r) - T_r(s - s_r)$$

(9)

where h is the specific enthalpy in kJ/Kg.

This equation was applied for all liquid streams in the power plant. For gas streams this equation was reorganised in terms of specific heat of gas c_p and evaluated as

$$a = c_p(T - T_r) - c_p T_r \ln (T / T_r)$$

(10)

where c_p is the specific heat in kJ/kgK.

The reference environment mentioned earlier was considered as temperature 27.8 °C and pressure 101 kPa. At this temperature and pressure, the enthalpy and entropy were obtained for water. Using Equation 9 exergy flow at different points of the power plant steam cycle was calculated. Equation 10 was applied to the different gas streams such as the air stream entering combustion and the flue gas following combustion.

The exergy flow rate can be calculated with the following equation

$$A = ma \qquad (11)$$

where A is exergy flow rate in kJ/s

The fuel enters the combustion at the reference environmental condition. Therefore, it enters the boiler combustor with only the chemical exergy with it. The chemical exergy flow of the fuel was calculated through Equation 12.

$$A = \sum_{i=1}^{i} \Psi_i^* \Phi_i \qquad (12)$$

where A in kJ/s, Ψ is Molar flow in kmol/s and Φ is Molar Exergy in kJ/kmol.

Standard Molar Exergy at 298 °C and 1 atm was found in the appendix of Moran and Shapiro [8] and used in Equation 12 to calculate exergy flow with fuel. The values of molar exergy of important chemical species in standard conditions are presented in Table 5. The reference temperature is slightly higher than this temperature. At this small temperature difference, the change of molar exergy is negligible and therefore ignored.

Components	Molar Exergy in (kJ/kmol)
Carbon	404590
Hydrogen	235250
Sulphur	598160
Oxygen	3950
Water Vapour	8635
Nitrogen	640

Table 6. Molar Exergy in Standard Condition

Once the exergy flows of all input and output streams were calculated, and the work inputs or outputs were obtained from different components, the destruction of exergy could be calculated by using Equation 13 after calculating all exergy inputs and outputs for a unit operation.

$$\sum_{i=1}^{i} A_i = \sum_{o=1}^{o} A_o + A_d \qquad (13)$$

where subscripts i, o and d are input, output and destruction.

The simulation of the whole process model produces process a mass and energy balance. The simulation performed the balance at steady state using the configuration inputs provided to each component described in section 3.3. There is a wide range of thermo-physical data available in addition to process mass and energy balances. Using those data the exergy flow calculation was performed at the different points. Equations 9~12 are used to calculate specific exergy and exergy flow at different points.

The balance of exergy flow against different components of the power plant was performed individually for each component observing exergy flows into and out of the component. Equation 13 is used to calculate an exergy balance which is similar to energy balance calculation described earlier. The same points in the power plant were used to observe the exergy flow of the streams as in the energy balance calculation. Similar to the energy flow calculation, the work inflow and outflow were obtained from the SysCAD process energy balance and were used in exergy balance calculations where they applied. The details of the exergy balance calculations of the power plant are presented in Table 7. As was the case for the energy balance calculation, the results of the exergy balance were exported to Microsoft Excel where, using equations described in Table 7, the exergy balance for all individual components was performed.

Components	Exergy Balance Equation
Boiler	$A_{26} + A_{27} + A_1 + A_3 = A_2 + A_4 + A_{28} + A_{30} + A_{d(Boiler)}$
HP Turbine	$A_2 = A_3 + A_{15} + W_{(HP\ Turbine)} + A_{d(HP\ Turbine)}$
IP Turbine	$A_4 = A_5 + A_{16} + A_{17} + W_{(IP\ Turbine)} + A_{d(IP\ Turbine)}$
LP Turbine	$A_5 = A_6 + A_{18} + A_{19} + A_{20} + W_{(IP\ Turbine)} + A_{d(LP\ Turbine)}$
Condenser	$A_{25} + A_6 = A_7 + A_{d(Condenser)}$
Condensate Pump	$A_7 + W_{(Condensate\ Pump)} = A_8 + A_{d(Condensate\ Pump)}$
LP1	$A_{24} + A_{20} + A_8 = A_9 + A_{25} + A_{d(LP1)}$
LP2	$A_{23} + A_{19} + A_9 = A_{10} + A_{24} + A_{d(LP2)}$
LP3	$A_{18} + A_{10} = A_{11} + A_{23} + A_{d(LP3)}$
Deaerator	$A_{22} + A_{17} + A_{11} = A_{12} + A_{d(Deaerator)}$
Feed Pump	$A_{12} + W_{(Feed\ Pump)} = A_{13} + A_{d(Feed\ Pump)}$
HP5	$A_{22} + A_{21} + A_{16} = A_{14} + A_{22} + A_{d(HP5)}$
HP6	$A_{14} + A_{15} = A_{21} + A_{d(HP6)}$

Table 7. Exergy Balance in Power Plant

In Table 7, the notation A represents exergy flow rate in kW. The subscript number in this table represents the point number in Figure 6. A with subscript d represents exergy destruction in the corresponding process mentioned within the bracket. W represents work and the corresponding process component is mentioned in a subscript within the bracket.

5. Result and discussion

As mentioned earlier, exergy analysis is performed to assess the loss of useful work potential; that is, the exergy destruction. The result of the exergy balance performed in the power plant is presented in Figure 7. In this figure, the exergy destruction is shown in percentages for different components of the power plant. The reason behind choosing percentage of total exergy destruction in different components is that these figures can easily be used to locate where the maximum exergy is destroyed. In other words, they can help direct the focus of the improvement by considering the components where most of the exergy is destroyed. The reason why this amount of exergy is destroyed is also important for identifying how best to reclaim the lost energy back in the process.

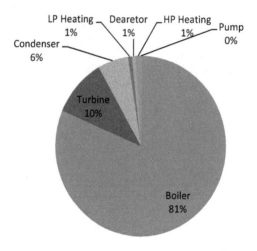

Figure 7. Exergy Destruction (%) in Power Plant

Figure 7 shows the result of this exergy destruction analysis. Most of the exergy is destroyed in the boiler, which accounts for 81% of total exergy destroyed in the power plant. It is important to note that the boiler includes both reheating of steam after expansion in the high-pressure turbine and preheating of air coming into the combustor. The second-largest source is the turbine where about 10% of the exergy is destroyed. These two components of the plant contribute more than 90% of the total exergy loss of the whole plant. The condenser is the third-largest contributor with about 6% of the exergy loss. The results of exergy analysis are markedly different from the results of the energy balance, which shows most of the energy being lost in the condenser.

A comparison of results of the energy and exergy balances in the power plant was conducted and the results are presented in Figure 8. It shows that there are very significant differences between exergy destruction and energy lost for different process components. It is important to mention here that only major components in the power plant, where most of energy or exergy loss or destruction take place, are considered in this comparison.

The energy balance showed that the primary source of energy loss is the condenser where 69% of the total loss occurs. In contrast, the exergy analysis showed that the loss from the condenser was only 6% of the total. According to the energy balance, the second-largest source of energy loss is the boiler, which accounts for 29%. However, the exergy balance revealed that the loss of useful work potential is in the boiler, with losses of more than 80%. It has been observed that there is a huge amount of energy lost in the condenser but the amount of useful energy, that is, exergy, is not very significant. In other words, it indicates that the waste heat in the condenser does not have much potential to be utilised as a source of work and to improve the efficiency the power plant. On the other hand, further investigation of the exergy lost in the boiler may show some opportunities for improvement.

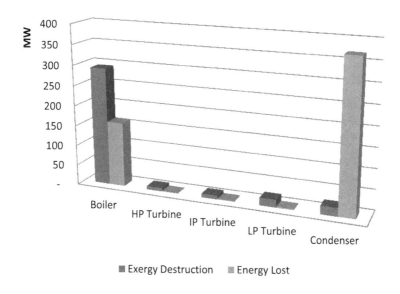

Figure 8. Exergy Destruction vs. Energy Loss in Power Plant

In an attempt to improve the efficiency of the power plant, the exergy analysis was revisited. It is found that the largest exergy destruction in the process occurs in the boiler. Therefore, it is considered first for a detailed investigation. It has been found that there are three elements which contribute to the huge boiler loss. They are 1) the boiler's internal heat transfer mechanism from the combustor to the heating medium, which determines the boiler's internal efficiency, 2) the heat loss in the departing flue gas stream and 3) the loss in the blowdown stream of the boiler. The contribution of the three losses of exergy flow in the boiler is presented in Figure 9. The simulation at steady state (280 MW electrical power output) showed that the flue gas after air preheating is leaving with an exergy flow of 4871 kW which constitutes about

2% of the total boiler exergy lost. The blowdown stream of the boiler is another source of boiler exergy destruction. It carries about 718 kW exergy flow. The boiler's internal heat transfer mechanism is responsible for most of the exergy loss occurs in the boiler.

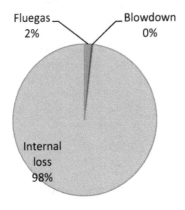

Figure 9. Boiler Exergy Losses in %

The exergy loses from flue gas and blowdown stream could be easily reutilised in the process through some heat recovery systems. However, the greatest single amount of exergy is destroyed in the boiler's internal heat transfer arrangement. If the exergy lost through the boiler could be utilised in the system, this would improve the efficiency of the power plant very significantly. The chemical reactions occur in the boiler at a temperature of 1885°C and produce huge amounts of exergy. However, due to the limitations of the material used in the boiler, it is not capable of transferring the full amount of useable heat energy or exergy to the boiler feed water which is heated to only 540°C. In modern power plants, this limitation has received much attention with the invention of new materials for heat transfer used in boilers. However, the power plant used as the subject of this study was an aging plant. The improvements needed to address such a big loss would require huge physical changes of the boiler system and may need further detailed investigation in terms of both technical and economic viability.

The exergy lost in the turbine is investigated by looking at the losses at different stages of the turbine. The exergy lost in all three turbine stages is found to be due to the turbine's internal performance. This can be better described as converting thermal energy to mechanical energy and then to electrical power. The turbine system used in the plant is highly compact and specially designed for the process. Therefore, the task of reducing exergy destruction or improving the efficiency of the turbine system is very specialised (in terms of its internal system). Similar to the boiler system improvement, it also needs further detailed investigation to assess the technical and economic feasibility of such changes.

It is noted that there are opportunities to improve energy efficiency of power plants by improving the performance of the boilers and the turbine system. The current trends towards ultra-supercritical power plant cycles are consistent with this aim.

6. Conclusions

In this study, exergy analysis of the power plant identifies areas where most of the useful energy is lost and discusses potential of the lost energy for improvement of the plant energy efficiency. It shows that the boiler of a subcritical power generation plant is the major source of useful energy lost. Only negligible amounts of useful waste energy can be recovered through implementing some heat recovery system. In order to achieve significant improvement of energy efficiency the boiler and turbine systems need to be altered, which require further techno-economic study.

Author details

R. Mahamud, M.M.K. Khan, M.G. Rasul and M.G. Leinster

*Address all correspondence to: r.mahamud@cqu.edu.au

Central Queensland University, School of Engineering and Built Environment, Rockhampton, Queensland, Australia

References

[1] SysCAD*SysCAD*. (2011). Available from: http://www.syscad.net/syscad.html.

[2] SysCAD*Introduction to creating a SysCAD project*, K.A.P. Ltd, Editor. (2010).

[3] IPCC*Climate Change 2007Mitigation. Contribution of Working Group III to the Fourth Assessment Report of the Intergovernmental Panel on Climate Change*, O.R.D. B. Metz, P.R. Bosch, R. Dave, L.A. Meyer (eds), Editor. 2007, Cambridge University Press: Cambridge, United Kingdom and New York, NY, USA.

[4] Ray, T. K, et al. *Exergy-based performance analysis for proper O&M decisions in a steam power plant.* Energy Conversion and Management, (2010). , 1333-1344.

[5] Regulagadda, P, Dincer, I, & Naterer, G. F. *Exergy analysis of a thermal power plant with measured boiler and turbine losses.* Applied Thermal Engineering, (2010). , 970-976.

[6] Som, S. K, & Datta, A. *Thermodynamic irreversibilities and exergy balance in combustion processes.* Progress in Energy and Combustion Science, (2008). , 351-376.

[7] Rosen, M. A. *Can exergy help us understand and address environmental concerns?* Exergy, An International Journal, (2002). , 214-217.

[8] Moran, M. J, & Shapiro, H. N. *Fundamentals of Engineering Thermodynamics.* 6 ed. (2008). John Willy &Sons Inc.

[9]   Bryngelsson, M, & Westermark, M. *CO2 capture pilot test at a pressurized coal fired CHP plant.* Energy Procedia, (2009). , 1403-1410.

[10] Korkmaz, Ö, Oeljeklaus, G, & Görner, K. *Analysis of retrofitting coal-fired power plants with carbon dioxide capture.* Energy Procedia, (2009). , 1289-1295.

[11] Dave, N, et al. *Impact of post combustion capture of CO2 on existing and new Australian coal-fired power plants.* Energy Procedia, (2011). , 2005-2019.

[12] Hammond, G. P. *Industrial energy analysis, thermodynamics and sustainability.* Applied Energy, (2007). , 675-700.

[13] Tsatsaronis, G. *Thermoeconomic analysis and optimization of energy systems.* Progress in Energy and Combustion Science, (1993). , 227-257.

[14] Amrollahi, Z, Ertesvåg, I. S, & Bolland, O. *Optimized process configurations of post-combustion CO2 capture for natural-gas-fired power plant-Exergy analysis.* International Journal of Greenhouse Gas Control, (2011). , 1393-1405.

Chapter 2

Oxy–Fuel Combustion in the Lab–Scale and Large–Scale Fuel–Fired Furnaces for Thermal Power Generations

Audai Hussein Al-Abbas and Jamal Naser

Additional information is available at the end of the chapter

1. Introduction

Recently, the environmental and health threat from anthropogenic emissions of greenhouse gases (GHG) of power plants has been considered as one of the main reasons for global climate change [1]. The undesirable increase in global temperature is very likely because of increase the concentrations of these syngas in the atmosphere. The most important resource of these anthropogenic GHG emissions in the atmosphere is carbon dioxide emissions. At present, fossil fuels provide approximately 85% of the world's demand of electric energy [2]. Many modern technologies in the electricity generation sector have been developed as sources of new and renewable energies. These new technologies include solar energy, wind energy, geothermal energy, and hydro energy. While these sources of renewable energy are often seen as having zero greenhouse gas emissions, the use of such technologies can be problematic. Firstly, sources of renewable energy are often still under development. Therefore, there can be a higher cost involved in their installation and in other related technical requirements. Secondly, the sudden switching of these energy sources (zero emission) has caused serious problems with the infrastructure of energy supply and global economy [3]. In order to reduce the problem and obey the new environmental and political legislation against global warming, it is necessary to find an appropriate solution to cut pollution which is with cost-effective, from the energy sources. The most effective technique, which can achieve a high level of reduction in GHG emission to atmospheric zone, is to capture carbon dioxide from the conventional power generations. At present, several organizations, energy research centres, companies, and universities, particularly in developed countries, are working to develop these conventional power plants in order to make them more environmentally friendly, with near-zero emissions sources. This chapter continues on different CO_2 capture technologies such as pre-combustion capture, post-combustion capture, and oxy-fuel combustion capture. The developments on oxy-fuel combustion technology with different scales of furnaces in terms of experimental investigations and theoretical modelling are discussed. The fundamentals and operating conditions of oxy-fuel-fired power plants are reviewed due to the importance of these conditions on the flame stability and coal combustion behaviour relative to those of conventional combustion. The effects of particular factors and parameters on the oxy-fuel combustion characteristics and boiler performance are reported. Finally, the chapter closes with a comprehensive CFD modelling study on the lab-scale and large-scale furnaces under air-fired and oxy-fuel combustion conditions.

2. Different CO_2 capture technologies

In order to understand the technologies that are used for CO_2 capture in the conventional power plants, it is important to understand the systems of leading technology for these power plants. The most popular leading technology systems are as follow: Integrated Gasification Combined Cycle (IGCC), Natural Gas Combined Cycle (NGCC) and Pulverised Fuel (PF) combustion steam cycles (some references called Pulverised Coal (PC)). As previously mentioned, large amounts of CO_2 emissions and other gases such as nitrogen oxides (NO_x) and sulphur oxides (SO_x) are produced by energy production from fossil fuel. Several techniques to capture carbon dioxide are being increasingly developed in order to comply with the new environmental and political legislation against global warming [3, 4]. The three main techniques, which have been developed for CO_2 capture from these different systems of leading technology, are pre-combustion capture, post-combustion capture, and capture of oxy-fuel combustion.

2.1. Pre-combustion capture

The pre-combustion capture technique excludes CO_2 from the fuel before the burning process in the combustion chamber. This technique can be achieved by installing special equipment which captures CO_2 between the gasifier and the combined cycle power plant. After gasification of coal or the reforming of natural gas with oxygen, the first step of this process leads to the production of a split stream of carbon monoxide (CO) and combustible gases (mainly hydrogen (H_2)). The second step is to convert CO to CO_2 with steam (synthetic gas with suitable amounts of water vapour) by a process called shift-conversion ($CO + H_2O \xrightarrow{yields} H_2 + CO_2$). After that, CO_2 can be separated by using a physical solvent, and finally the CO_2 becomes efficiently available for a sequestration process after it passes through a compression unit. The other remaining parts (mainly hydrogen (H_2)) will be sent to the combined cycle power plant to be used as input fuel for power production [5]. As shown in Figure 1, the clean syngas is supplied to a combined cycle power plant after several treatment processes such as gas cooling (for protecting equipment), particulate removal, and hydrogen sulphide (H_2S) removal. Although this method of CO_2 capture can be considered a good producer of hydrogen for the combined cycle power plant, it has a relatively high level of complexity compared with other CO_2 capture techniques [6].

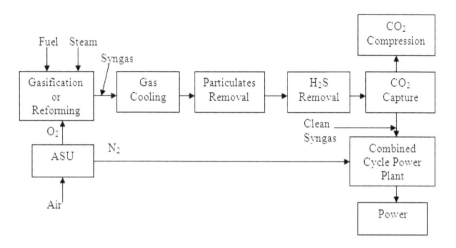

Figure 1. Power plant with pre-combustion capture technology

2.2. Post–combustion capture

The post-combustion capture technique involves capturing CO_2, as well as reducing particulate matter, SO_x, and NO_x in the combustion flue gases (see Figure 2). This technique requires adding a separation unit after firing systems of the PC or NGCC. Any of the following three separating technologies can achieve the sequestration of CO_2: chemical absorption, low temperature distillation, and gas separation membranes [7].

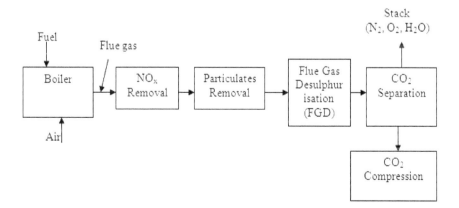

Figure 2. Power plant with Post-combustion capture technology

In the first separating technology, the chemical absorption of chemicals such as monoetha-nolamine (MEA) is used in an absorption tower to scrub CO_2 from the combustion flue gases. A high temperature level is used to separate CO_2 from the chemical solvent after delivering it to the regenerating tower. After that, a compression process is performed to capture CO_2 as seen in Figure 3. Because of the relatively high temperature and low partial pressure (concentration) of the carbon dioxide to be treated in the flue gases, this method offers a big design challenge for conventional power generation. This means that the chemical absorption process can only provide an economic benefit if it is applied to work at a small scale. On the other hand, this process needs a large amount of energy due to the large size of the main sequestration components, and thereby the energy penalty will introduce a higher operating cost if it is applied for large-scale power plant with CO_2 capture [8, 9]. The second separating technology for post-combustion capture uses gas separation membranes such as solution-diffusion or molecular sieving, which can be used to capture CO_2 by separating it from the flue gases. These membranes can experience some technical problems if applied to the capture of CO_2 in flue gases from coal power plants due to the degradation of the absorbent by impurities existing in the flue gas. However, this technology has not yet shown its ability to be applied at a large-scale CO_2 capture power plant, and it is still under development. Finally, the third separation technology, low temperature distillation, can be used to capture CO_2 from the flue gas, but this process requires special conditions (above 75 psi pressure, and -75 °F temperature) to achieve a high purity of CO_2 (about 90% CO_2) in the flue gas. Due to these complicated conditions involved in the separating processes in a power plant, low temperature distillation is not considered an efficient technology for the CO_2 capture from power plants.

2.3. Oxy–fuel combustion capture

The oxy-fuel combustion technique captures carbon dioxide from the flue gases of combustion. It is approximately similar to the post-combustion capture technique in terms of separating the CO_2 from the exhaust gases as a final process of sequestration, but it is less chemically complicated. As described earlier, the partial pressure of CO_2 in the flue gas is low in conven-tional combustion (air-fired combustion), and it needs special treatments for the separation processes. The basic principle of oxy-fuel combustion is to increase the partial pressure of CO_2 in the flue gases in order to make its sequestration and compression process easier and more cost-effective. This technique can be performed by using a mixture of pure oxygen with part of a recycled flue gas (RFG) (mainly CO_2) instead of air in the combustion chamber. In this case of combustion, a high concentration of CO_2 (high partial pressure) can be achieved in the flue gas stream, and therefore the high cost of its capturing processes can be avoided unlike the post-combustion process. The oxy-fuel combustion technique is schematically shown in Figure 3. In Figure 3, the oxy-fuel technique shows that the first removal equipments extract particulates and sulphur dioxide, respectively. The particulate removal can remove the fly ash from the flue gas, while the bottom ash is removed after accumulating at the bottom of the furnace, i.e. at the hopper zone.

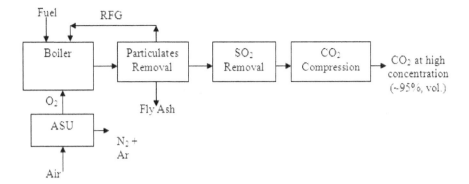

Figure 3. Power plant with the capture of oxy-combustion technology

After removing particulates and condensing water vapour from the flue gases, the concentration of CO_2 will be increased to around 75% vol. under wet basis or to around 95% vol. under dry basis so that it can be transported for permanent storage. The purity of CO_2 concentration is completely dependent on factors such as the purity of the oxygen feed (from ASU), air-leakage, and the excess of the oxygen/fuel (stoichiometry) ratio. As shown in Figure 3, a part of flue gas (around 60-70%) is recycled and mixed with pure oxygen. This process is used to prevent high temperature levels inside the furnace, i.e. to protect the furnace wall. Due to this recirculating of flue gases, the size of the furnace and the sizes of other gaseous removal equipment can be significantly reduced. The recycled gas process of oxy-fuel combustion technique can maintain the same flow field conditions of the burners in the conventional combustion case.

The air separation unit (oxygen production unit) produces two gas streams. The first is an oxygen stream, which supplies the furnace, and the second includes nitrogen and other minor constituent inert gases that are vented to the atmosphere. In the air separation unit (ASU), a large amount of energy is used to separate nitrogen and produce a pure stream of oxygen. Despite the high cost of this process, oxy-fuel combustion is definitely a competitive technique compared with the other CO_2 capture technologies due to the high reduction in NO_x and SO_x emissions besides its high CO_2 capture. However, using pure oxygen and recycled flue gas (RFG) instead of air to burn pulverized coal (PC) in the combustion chamber leads to problems such as many changes in flame temperature levels, species concentrations, and radiation heat transfer problems inside the furnace. These modifications of the combustion characteristics are due to the following reasons: the high specific heat capacity of CO_2 with respect to nitrogen in conventional combustion, radiative properties of gas mixtures, low oxygen molecular diffusivity in CO_2 compared to N_2 and other transport properties of the gas mixture such as viscosity, thermal diffusivity, gas phase chemistry etc. [1, 3, 10-12].

2.4. Comparison amongst different capture technologies

In short, all these three CO_2 capture technologies have different outcomes, particularly with regards to reduction of power plant efficiency and in increasing the cost of electricity production. In general, to be successful post-combustion capture requires new developments in the process of chemical absorption of CO_2 in order to adequately reduce energy consumption in the absorption process, but this is very expensive. In contrast, pre-combustion capture is achieved by the conversion of fuel into carbon monoxide (CO) and hydrogen fuel (H_2) in which CO is converted to CO_2 by the shift-conversion process. This CO_2 capture approach can be developed by either physical or chemical absorption processes to avoid any extra complexity in chemical design of power generation. However, both of these processes are very expensive and chemically complicated. Finally, capture of CO_2 by the oxy-fuel combustion technique is less expensive than the other two processes and less complex. It can be carried out by burning the fossil fuel with a mixture of pure oxygen (99.5 vol. %), produced in ASU, and recycled flue gas (RFG). The products of this combustion will be only CO_2 and H_2O in the flue gas. After the condensation process, CO_2 concentrations will be increased to a level more suitable for the separating and compression processes. Although energy consumption for O_2 production, in an air separation unit, is relatively high (about 10% of the net energy of power plant), new technologies for air separation processes and O_2 production are under development in order to reduce the energy penalty for CO_2 capture power plants [13].

Recently, Kanniche et al. (2009) [7] comprehensively made a comparison between the leading technology systems for the IGCC, NGCC, and PC power plants with the above-mentioned CO_2 capture technologies. The results showed that the efficiency of the PC power plant when it used post-combustion capture is lower than that of the IGCC power plant using pre-combustion capture. The NGCC and PC power plants obtained the highest efficiencies with post-combustion capture and with oxy-fuel combustion, respectively. Regarding the production costs, the lowest costs occurred when PC used oxy-fuel capture, but costs gradually increased for IGCC use of pre-combustion capture and NGCC use of post-combustion capture. The highest cost per tonne of CO_2 removal was for NGCC using pre-combustion capture, whereas the costs for PC using oxy-fuel combustion capture and PC using post-combustion capture were approximately at the same cost levels. Based on these results, Kanniche et al. recommended taking the following considerations into account during designing near-zero emissions power plants: pre-combustion capture in IGCC, post-combustion in NGCC, and oxy-fuel combustion in PC.

In addition, oxy-fuel combustion is a considerably competitive technique compared to other CO_2 capture technologies. Therefore, a large number of experimental and theoretical studies are being carried out in this area worldwide, particularly in developed countries using different scales of furnaces. Generally, these studies depend on solid fuel (coal) because it is a major source of energy in some of these countries. In order to identify any difficulties in using a large-scale boiler during switching from air-fired (conventional) combustion to oxy-fuel combustion, several studies using different scales of furnaces should be widely undertaken. These studies on oxy-fuel-fired scenarios will provide relevant information to maintain the similar combustion characteristics of PC without capture. Thereby, the most cost-effective basis can be achieved for retrofitting existing power plants or to build a new power plant unit under oxy-fuel combustion conditions. However, recent studies have concluded that the oxy-fuel combustion technology is technically feasible and can be applied in a large-scale pulverized coal-fired power plant, and it certainly represents a competitive method relative to the other CO_2 capture technologies. As a result, investigations, developments, innovations and research in this technique are necessary to provide high level confidence and operational experience at a laboratory-scale and pilot-scale and then gradually at a large power plant scale.

3. Oxy–fuel combustion technology developments

In order to reduce GHG emissions to the atmosphere, systematic research and development work on the existing electricity power plants for CO_2 capture are required. Research into CO_2 capture started two decades ago with both experimental investigations and numerical simulation methods. Lab-scale furnaces were initially used in order for researchers to identify and characterise the fundamentals and operations of oxy-fuel combustion issues under different operating conditions. The fundamental aspects of concern included, for example, flame stability [14, 15], ignition behaviours [16, 17], species concentrations [18], and fuel combustion rate [19]. The heat transfer characteristics under different oxy-fuel-fired scenarios have been studied [20, 21] in order to reduce the retrofits needed to convert conventional boiler designs to oxy-combustion processes. Globally, there have been some studies on pilot-scale oxy-fuel combustion facilities in Europe and some developed countries [22-24]. These investigations revealed that power plants can simply switch from air-fired combustion to the oxy-firing at a large scale and produce higher concentrations of CO_2 in the flue gas. In addition, a significant reduction in NO_x emissions can be achieved due to eliminating N_2 from the oxy-fuel combustion processes [25]. These confirmations and support for oxy-fuel combustion technique for CO_2 capture show that there are no main crucial barriers in implementing this clean, efficient, and economic technology in industrial large-scale facilities. However, switching to oxy-combustion is completely dependent on public support and government developing policies to address global climate change.

3.1. Experimental and theoretical laboratory–scale projects

With increasing concerns from the Kyoto Protocol about the CO_2 emissions and global warming, research into oxy-fuel combustion technology has widely increased. To overcome the difficulties in applying the oxy-fuel combustion and make it more appropriate and acceptable in the applicable fields, several experimental studies have been conducted with different combustor sizes such as lab-scale and pilot-scale furnaces as the basis for building a large-scale oxy-fuel furnace. In the literature on oxy-fuel combustion, studies on lab-scale furnaces have mainly focused on some important points: ignition behaviour, chemical species characteristics, char combustion, flame propagation speed, NO_x reduction and SO_x formation, and heat transfer models were investigated.

Liu et al. (2005) [19] used a 20 kW down-fired coal combustor (190 mm inner diameter and 3 m height) to test the UK bituminous coal combustion in air and in the mixture of O_2/CO_2. The authors showed that the char burnouts and gas temperatures obviously decreased in coal-O_2/CO_2 combustion due to the high specific heat capacity of carbon dioxide compared to nitrogen. They recommended that the concentration of oxygen in O_2/CO_2 mixture should be increased to 30% and 70% for CO_2 (or recycled flue gas) to achieve a corresponding temperature similar to the coal-air combustion.

The influence of reactions, including the effects of char structure and heat transfer, on the ignition behaviour of low-rank Victorian brown coal and high-rank Chinese bituminous coal in air-fired and oxy-fuel combustion cases was experimentally investigated in a wire-mesh reactor by Qiao et al. (2010) [26]. As the gas mixture was set at 21% O_2 and 79% CO_2, a slight increase in the average ignition temperature was noted for both the coal types. In contrast, a noticeable decrease in the particle ignition temperature was observed with increasing O_2 concentrations for brown coal and bituminous coal during oxy-fuel-fired scenario. Qiao et al. concluded that the reason for that ignition behaviour was the thermal physical properties of the gases surrounding the particles (Qiao et al. 2010). As a continuation of the Victorian brown coal investigation into O_2/N_2 and O_2/CO_2 mixtures, Zhang and co-researchers (2010) [17] and (2010) [27] used a lab-scale drop-tube furnace (DTF). The measurements were conducted using a high-speed camera (MotionPro Y-3) and two-colour pyrometers, the first for photographic observation, and the second for particle temperature measurements. The authors concluded that the coal pyrolysis, coal combustion (volatiles ignition and char oxidation rate), and the surface temperature were highly influenced by the bulk gases in the O_2/CO_2 mixture. There was a clear delay in the coal ignition in the oxy-fuel combustion environment compared to that in the coal-air combustion. This was because a thick volatile cloud released remained attached to the char surface for a long time in the O_2/CO_2 mixture and led, as a result, to the high oxygen consumption on the char surface. Zhang et al. recommended increasing the O_2 concentration in the O_2/CO_2 mixture to 27% and 73% for CO_2. This increase leads to achieving good stability in the volatile flame, and obtaining a corresponding char particle temperature to that of the air-fired case.

Computational fluid dynamics (CFD) modelling studies can comprehensively provide a wide range of information for the design of furnace and burner that can reduce the cost of time-consuming experimental investigations. The CFD has the ability to predict well the flame structure, gas temperatures distributions, chemical species concentrations, radiative heat transfer etc., under different combustion conditions. One of the CFD benefits is that the multiple chemistry mechanisms discovered can be used in simulations of the fuel reaction, which are often conducted on the assumption of a chemical balance, finite-rate chemistry scheme, or an approach of mixed-is-burned model [28, 29].

Recently, Venuturumilli and Chen (2009) [30] performed a CFD analysis between the four-step reduced mechanism and its starting mechanism on axisymmetric laminar diffusion (non-premixed) methane flames. The reduced mechanism included 8 major species (CH_4, H, H_2, H_2O, CO, CO_2, O_2, and N_2), while the starting mechanism had 18 species and 65 elementary reactions. The comparison between these two different chemical mechanisms showed that the temperature distributions and axial velocity profiles were similar at the flame base location, whereas the four-step reaction mechanism was not able to provide precise information about the ignition characteristics of the methane flame. Furthermore, the authors estimated the computational time required for the starting mechanism was around 3-4 times as long as for the four-step reaction mechanism. In addition to the above experimental investigations and numerical modelling, the open literature includes a number of relevant findings [31-33] from small lab-scale oxy-fuel furnace experiments. These lab-scale studies under oxy-fuel combustion conditions are useful, and can provide some technical insights and fundamental engineering techniques for this challenging technology. The uncertainties about the heat transfer characteristics, flame behaviour, corrosion problems and pollutant control units in lab-scale furnaces lead to the need for further research on the application of this technology prior to industrial full-scale boiler development.

3.2. Industrial large–scale demonstration developments

Developments in the lab-scale oxy-fuel facilities as a result of experimental investigations and numerical simulation methods have led to new approaches being used in large-scale facility units. These developments have allowed, effectively, a good compromise between the expensive experimental tests and complete simulation of commercial large-scale with CCS plants. Therefore, the knowledge and technical experience gained from these studies and investigations can be utilized directly to design industrial demonstration large-scale boilers (>250 MW) for the next few years. However, the pollutant control units are not completely understood in oxy-fuel combustion systems, and as a result more research work is required, particularly in the mercury removal units. A lot of information in regard to the international demonstration of oxy-fuel combustion can be found in the International Energy Agency (IEA) Greenhouse Gas Programme, which has been supported by the Asia partnership project. Recently, there have been a number of projects investigating PC oxy-fuel combustion demonstrations, some focussing on electricity production and other on CCS processes in industrial large-scale boilers such as Vattenfall, Endesa, FutureGen 2.0, and KOSEP/KEPRI, as summarized in Table 1 [34]. The implementation of these industrial-scale oxy-fuel combustion projects will greatly reduce the technology costs, especially in the ASU and CO_2 compression equipment, and make it more appropriate for commercial applications in 2022, as reported in the sequence project developments by Wall *et al.* (2009) [1].

Up to date, in the field of numerical simulation on the commercial-scale facility unit, there has been unfortunately little research work conducted on oxy-fuel combustion conditions. Zhou and Moyeda (2010) [35] conducted a process analysis and main calculations on an 820 MW facility to compare the furnace temperature profiles for the air-fired and the oxy-fuel combustion with both wet and dry recycled flue gas modes. The authors proposed several design criteria for both oxy-fuel combustion cases as follows: they would use the same total heat fuel, the same boiler exit O_2 and the same gas flow rate in order to reduce the retrofit impact on the conventional boiler performance. They showed that the flue gas recycle ratio depends on the stoichiometric (air-to-fuel) ratio and ash content in coal. Their results also indicated that the moisture content in O_2/CO_2 combustion was 3.5 times higher than that of the conventional combustion. As a result, a clear reduction in the adiabatic flame temperature was noticed due to the increased moisture content, and therefore 75% of the recycled dry flue gas should be used to increase the flame temperature. The significant resulting increase in the H_2O and CO_2 concentrations in the flue gas of O_2/CO_2 combustion cases was accompanied by a clear decrease in NO_x formation due to the reduction in the flame temperature and thermal NO production.

Project	Location	MW$_{th}$	Start Up Year	Boiler Type	Main Fuel
B & W	USA	30	2007	Pilot PC	Bit, Sub B, Lig.
Jupiter	USA	20	2007	Industrial	NG, Coal
Oxy-coal UK	UK	40	2009	Pilot PC	Bituminous
Alstom	USA	15	2009	Pilot PC	Bit, Sub B, Lig.
Vattenfall	Germany	30	2008	Pilot PC	Lignite (Bit)
Total, lacq	France	30	2009	Industrial boiler	NG, Coal
Callide	Australia	90	2011	PC-with Electricity	Bituminous (Sub B)
CIUDEN-PC	Spain	20	2010	Pilot PC	Anthra, Bit, Lig.
CIUDEN-CFB	Spain	30	2011	Pilot CFB	Anthra, Bit, Lig.
ENEL HP Oxy	Italy	48	2012	Pilot-High pressure	Coal
Vattenfall	Germany	1000	2014	PC-with Electricity	Lignite (Bit)
Endesa	Spain	1000	2015	CFB-with Electricity	Anthra, Bit, Lig.
FutureGen 2.0	USA	600	2017	PC-with Electricity	Coal
KOSEP/KEPRI	Korea	400	2018	PC-with Electricity	Coal

Table 1. Developments of oxy-fuel combustion demonstration projects around the world, adopted from [34]

4. Fundamentals and operations of oxy–fuel combustion power plants

In this section, some of the fundamental issues affecting the operations of oxy-fuel power plant, e.g. coal ignition, flame stability, and char combustion are briefly surveyed. For better understanding of these aspects of the combustion processes, the survey includes both the experimental results and theoretical methods. Other important issues that will be reviewed include the effects some parameters, such as oxygen content, particle size, RFG ratios, and air leakages have on the combustion characteristics. These parameters are considered to have a significant influence on the performance and reliability of oxy-fuel power plants.

4.1. Coal ignition

To understand the impact of the specific heat capacity of the gas mixture on the coal ignition process, the following mathematical equations can be inferred from the above observations.

Once the volatile matter has been released during the volatilization process, the auto ignition time of coal particles can be determined based on the assumptions of ignition and explosion theory [36]. For a one-step overall reaction and no heat loss, the ignition delay of volatile gases is given as follows:

$$\tau = \frac{c_v}{Q_c Y_{F,O} A \exp(-T_a/T_o)} \frac{T_o^2}{T_a} \tag{1}$$

Where, c_v is a specific heat capacity at constant volume, Q_c is the combustion heat release per mass of fuel, $Y_{F,O}$ is the mass fraction of fuel at initial time value t=0, T_o is the initial temperature of reactants, T_a is the ambient temperature and A is a kinetic factor in the Arrhenius expression. According to Eq. 1, it is clear that the ignition delay time increases with an increase in the specific heat capacity (c_v) of gases and decreases in proportion to the combustion heat release (Q_c). Therefore, the high specific heat capacity can play an important role in increasing the ignition delay. As mentioned earlier, the specific heat capacities of the gaseous combustion products such as CO_2 and H_2O are higher than that of N_2. As reported in many oxy-fuel experimental investigations [16, 37, 38], the ignition delay can be reduced by increasing the oxygen concentration in the gas mixture in order to enhance the chemical reaction rate, as explained in Eq. 1. As a result, a higher O_2 concentration can be used to dilute the carbon dioxide effect on the ignition mechanism and to yield the same ignition time for oxy-fuel combustion as that of coal-firing in the air.

In addition, the chemical effects on the coal particles, due to the elevated concentrations of CO_2 and H_2O in the flue gas of oxy-fuel combustion, have been considered as another reason for the coal ignition delay. This chemical phenomenon was observed in two experimental studies, which were conducted in lab-scale drop-tube furnaces (DTF) in Australia [39, 17]. The authors used different types of Australian coals and a wide range of O_2 levels to investigate the coal pyrolysis behaviour, ignition extent, and char burnout in air-fired and oxy-fuel-fired environments. Zhang et al. (2010) [17] noted that when the nitrogen is replaced by carbon dioxide in the gas mixture, this enhances coal pyrolysis prior to ignition and as a result produces a large cloud of thick volatile matter on the char surface. Thereby, ignition of the volatile cloud occurred instead of single particle ignition. Approximately 30% of O_2 concentration in the O_2/CO_2 mixture was recommended to be used in order to achieve similar coal ignition like that in the O_2/N_2 mixture. Rathnam et al. showed that the apparent volatile yield measured in the DTF for O_2/CO_2 mixture at 1673 k was around 10%, as high as that in air-firing for all coal types.

4.2.Flame behaviour

In general, the physical parameters of the flame, for instance shape, brightness, extinction, oscillation frequency, and temperature level, can be used to characterise the flame behaviour and control the stability of flame in the furnace to improve combustion efficiency. As a consequence, flame stability is an important issue and needs to be studied and taken into consideration when designing burners in order to improve flame characteristics and reduce emission levels. The replacement of air by CO_2 in the feed oxidizer gases has been found to have a significant effect on flame stability. However, to maintain a better flame stability in an oxy-fuel furnace, Chui et al. (2003) [22] tested two different burners' configurations (A and B) to compare which has optimal efficiency. Both experimental investigation and numerical modelling studies on a pilot-scale furnace were conducted. The main difference between the designs of the two burners was the location of the pure oxygen injection into the primary gas stream. For burner A, the annulus high-O_2 jet was located inside the primary stream, while in burner B, the annulus high-O_2 jet was located between the primary and secondary streams and without a cyclone chamber for coal delivery. The experimental and numerical results showed that burner A improved the flame stability and achieved a significant decrease in the NO_x level in the combustion flue gas in comparison with burner B, particularly when the swirl number was increased in the secondary stream. The improvement was due to the enhancement of the internal recirculation zone of reactants in the near-burner region.

Using state-of-the-art turbulence-chemistry interaction, several advanced turbulent combustion models have been developed such as Eddy-Breakup (EBU) model, Probability Density Function (PDF) transport model, and Conditional Moment Closure (CMC). Most of these mathematical combustion models have only been utilized for the calculations of air-fired flame. For that reason, Kim et al. (2009) [40] adopted the CMC model to analyse the characteristics of turbulent combustion of natural gas (NG) flame in air-fired and oxy-fuel combustion environments. The authors coupled the CMC model with a flow solver. The detailed chemical kinetics mechanism model was implemented to calculate the intermediate species (CO and H_2) formed in the flame envelope because of the enhanced thermal dissociation. The normalized enthalpy loss variable (ξ) used to calculate the effect of convective and radiative cooling terms on the turbulent flame, and it is defined as follows:

$$\xi = h - h_{min} / h_{ad} - h_{min} \tag{2}$$

Where, h_{ad} is the conditional adiabatic enthalpy, and h_{min} is the conditional minimum that is calculated when the conditional temperatures are reduced to the surrounding temperature.

The numerical results showed that the oxy-fuel flame produced a much broader region in the hot-flame zone, particularly in the lean-fuel side of the mixture fraction, in comparison with the air-NG flame. They also noted that the temperature value of oxy-fuel flame is higher than that of air-firing. The predicted temperature levels for both combustion cases were overestimated compared to the experiments. That discrepancy arose because of the inadequacy of the modified $K - \varepsilon$ turbulence model used in the calculations and to the measurements uncertainties. In order to mathematically demonstrate the chemical effect of CO_2 on the reduction of propagation flame speed, the extinction theory of diffusion and non-diffusion flames will be considered. The Damkohler number (D) is a dimensionless number, and can be used to interpret the extinction characteristics of flames, as follows [41, 42]:

$$D = \frac{\tau_r}{\tau_{ch}} \tag{3}$$

Where τ_r represents the residence timescale of reactants in the combustion zone, also known as the mixing timescale, and τ_{ch} is the chemical reaction timescale and it is equal to a ratio of the characteristic thermal diffusivity of the gas divided by the square laminar burning speed (i.e. $\tau_{ch} = \alpha / S_L^2$). The numerator of the D number is completely dependent on the fluid dynamics of the flame, but the denominator of D is a function of the flame reaction rate. Therefore, the mathematical expression of the Damkohler number can be rewritten using the Arrhenius kinetic rate formula, as follows:

$$D \propto \frac{\tau_r}{\alpha} \; exp(-E_A / R \, T_f) \qquad (4)$$

According to Eqs. 3 and 4, flame velocities in oxy-fuel combustion environments will observably be lower than those in air-fired environment at the same level of oxygen concentration. This is due to the lower values of thermal diffusivity and adiabatic flame temperature in O_2/CO_2 mixtures compared to those in O_2/N_2 mixture. As a result, the flame stability of the oxy-fuel combustion will definitely be affected by that reduction attained in propagation flame speed.

4.3. Char combustion

After the devolatilization process of the coal particles finishes, the char combustion subsequently starts in the firing system. This combustion process has been considered the dominant factor constraining several reaction parameters such as the total burnout time, unburned carbon level, and radiation from burning char particles. A better understanding of the effects of these parameters on the combustion characteristics and boiler heat transfer under oxy-fuel conditions enables engineers optimize the applications of both the existing and new coal power plants.

In order to reduce the char burnout time and increase the combustion rate of char particles in O_2/CO_2 mixtures, oxygen-enriched atmosphere must be used in the gas mixture [23]. However, the literature shows that the combustion rate affects the reaction order of the bulk oxygen partial pressure. This means that the combustion rate when measured in terms of the chemical kinetic control reaches a high reaction order (0.6-1) for O_2 below 900K, but a low reaction order in the oxygen partial pressure is exhibited when the global reaction rate is above 1200K [43]. Based on these findings, the reaction rate of coal char particles has been interpreted to be subject to an n-th order law governing oxygen partial pressure and the Arrhenius kinetic rate model of char reaction. The overall rate of gasification is given as follows:

$$r_{gas} = k_s(T_p) \, P_{O_{2,s}}^n \qquad (5)$$

Where n represents the reaction order, T_p is the temperature of the coal particle, and k_s is a coefficient of temperature dependent rate and can be written according to the Arrhenius expression: $k_s(T_p) = A \, exp(-E / R \, T_p)$.

The elevated CO_2 and H_2O concentrations in the flue gas of oxy-fuel combustion will influence the reaction rate of char particles. As previously mentioned, this can be attributed to the relatively high specific heat capacities of the main gaseous products (CO_2 and H_2O) compared to the nitrogen in the air-firing. These dominant species will act with the remaining char particles following the endothermic process and thereby reduce the char particle temperature, resulting in a decrease in the rate of char oxidation. Furthermore, the diffusivity of oxygen on the char surface in an O_2/CO_2 mixture is lower than that in an O_2/N_2 atmosphere, and this could be why a lower char oxidation rate is observed in an O_2/CO_2 mixture. Due to the importance of the elevated gases on the gas temperature and char burning rate, Hecht *et al.* (2010) [44] used a computer model to study the effect of the endothermic CO_2 gasification reaction on the char consumption under different oxygen-enriched environments. Numerical modelling was implemented employing the Surface Kinetics in Porous Particles (SKIPPY) code over a range of potential CO_2 oxidation rates for bituminous coal particles. The SKIPPY depends on the FORTRAN program to solve the conservation mass, momentum, energy, and species concentration equations by assuming a multicomponent gaseous phase. The results showed that where there was 12% oxygen in an O_2/CO_2 mixture, the endothermicity of the CO_2 gasification led to a significant decrease in the char particle temperature, and thus reduced the reaction rate of char oxidation. For up to 24% O_2, the global consumption rate of char particles enhanced with increasing the reaction rate of CO_2 gasification. With more than 24% O_2, the overall rate of char reaction decreased with an increase in the gasification rate of carbon dioxide.

4.4. Effect of some parameters on oxy–fuel characteristics

When a mixture of pure oxygen and recycled flue gas is used instead of air as the combustion gases in oxy-fuel-fired power plants, many modifications to the combustion characteristics of the power plant boiler will occur under normal operating conditions. Therefore, in order to address these challenges and ensure the oxy-fuel power plants are working at high combustion efficiency, consistent thermal performance, and with low emissions, several major parameters can be utilized to achieve that remarkable goal. Based on the design and operating conditions of the existing power plants, the effects of different parameters under oxy-fuel combustion conditions such as oxygen concentration, particle size, and recirculation of flue gas have been investigated. In this subsection, a brief summary of research under the effects of these different parameters is presented.

4.4.1. O_2 concentration

The oxygen concentrations in the mixed oxidizer stream have a significant impact on the flame stability and heat transfer characteristics in the oxy-fuel firing facilities. In order to maintain the same aforementioned characteristics as in the conventional firing systems, the desirable value of the O_2 concentration has to be precisely determined. Due to the physical and chemical differences between the properties of carbon dioxide and nitrogen, it seems that 21% O_2 concentration (by volume), in a mixed stream of oxy-fuel conditions, does not provide the same combustion characteristics as conventional operations. This can be explained, as mentioned earlier, as a result of the delay in the ignition time of volatile matter released and to the difficulties in oxidation of the coal char particles. Therefore, higher oxygen concentrations are required to work safety and provide more efficient operations. However, the RFG enriched with 28 vol. % O_2 is safer than of 21% O_2 in the secondary stream of a pilot-scale burner in oxy-fuel combustion tests for high-volatile subbituminous coal [22].

The oxy-fuel combustion scenarios offer the opportunity to supply different amounts and concentrations of RFG enriched with O_2 to both the fuel carrier gas and feed oxidizer gas streams. Many studies have shown that a range between 25% and 36% oxygen (by volume), at the burner inlet, is preferable in oxy-fuel-fired conditions to maintain the same flame behaviour and heat transfer characteristics as those in the air-fired [14, 17, 45-47]. In addition to the safety reasons, this range of oxygen content in the gas mixture was basically found to result in a lower level of pollutants in the flue gas such as NO_x, SO_x, CO, and trace elements during the burning of several coals types. By increasing the O_2 concentrations to more than 21 vol. % in both the primary transport gas and RFG streams at the commercial low NO_x swirl burner, Sturgeon et al. (2009) [33] noted some improvements were achieved in different areas of investigation compared to the standard conventional firing. For example, the level of CO concentration was decreased, a stable flame was attained, and a low content of carbon in the ash was noticed.

4.4.2. Particle size

The findings on the effect of coal particle diameters on the oxy-coal firing have shown that there are recognizable influences on the flame propagation speed, devolatilization process, and ignition temperature. Suda et al. (2007) [47] used two different particle diameters (50 μm and 100 μm) to investigate the effect of coal particle size on the propagation behaviour of flames. The results showed that the flame propagation velocity slightly decreased with coal particle size. However, the reason for that decrease is the lower heat transfer conduction process between the coal particles and gas. Moreover, the authors concluded that the flame stability in the O_2/CO_2 mixture could be increased using smaller PC particles.

Huang and co-researchers (2008) [46] investigated the effect of the coal particle size on the combustion characteristics of TieFa (Tf) bituminous coal in different mixtures of O_2/CO_2 atmospheres during the experimental tests. The experiments were conducted using a differential thermal analyser (DTA). Three different particle sizes (11.34, 18.95, and 33.68 μm) and four different O_2 concentrations (10, 20, 50, and 80%) in the gas mixture were used in order to carry out comprehensive investigations into the effect of these two important parameters. The authors used a ratio of thermal gravimetric (TG) to the differential thermal gravimetric (DTG) to clarify the combined effect of the above-mentioned factors on both the devolatilization and char combustion processes. The results of the TG/DTG curves showed that the weight loss of coal samples was augmented as the coal particle size decreased. That contributed to an increase in the surface area of coal particles that led, as a result, to enhance the overall reactivity of coal char, particularly when the O_2 concentration was increased from 10% to 80% at a constant heating rate. In addition, there was a clear decrease in both the ignition and burnout temperatures at the small coal particle size. The effect of the particle size on the reaction rate was relatively negligible at oxygen-rich conditions (50% and 80%). The combustion property index

(S) was also used to give an inclusive estimation of the combustion characteristics under these specified conditions and practical operations. The index (S) was defined as the coal reaction rates divided by the square of ignition temperature and burnout temperature, as follows [48]:

$$S = \frac{\left(\frac{dW}{dt}\right)^c_{max} \left(\frac{dW}{dt}\right)^c_{mean}}{T_i^2 \, T_b} \qquad (6)$$

Where $\left(\frac{dW}{dt}\right)^c_{max}$ and $\left(\frac{dW}{dt}\right)^c_{mean}$ represent the coal burning rates at the maximum and mean values, respectively. The S values were determined with a constant heating rate of 30 °C/min. The plotting of the combustion property index against the coal particle sizes show that the index S not only increased with a decrease in the particle size, but also with an increase in the oxygen concentration. Consequently, they concluded that the S values become superior with these two intensified combustion parameters [46].

4.4.3. Flue gas recirculation

The recirculation of flue gas or recycled flue gas (RFG) is an important process in oxy-fuel combustion. In this process, a large amount of combustion flue gases is recycled to the furnace in order to maintain the same combustion temperature and heat transfer characteristics in the boiler as in the conventional coal-fired power plant. The recycle ratio of flue gas can be defined according to the following simple mathematical expression [49]:

$$Recycle \ ratio = \left(\frac{Recycled \ gas \ mass \ flow \ rate}{Recycled \ gas \ mass \ flow \ rate + Product \ gas \ mass \ flow \ rate} \right) * 100 \qquad (7)$$

Typically, 60% to 80% of produced flue gas, mainly CO_2 and H_2O, is recycled in the oxy-fuel plants, and this ratio is basically dependent on the coal type and the options of RFG [35]. As previously illustrated in Fig. 4, the oxy-fuel technique offers two options to draw the flue gas, either wet RFG or dry RFG can be practically used for the same purpose, depending on from which locations the flue gas is taken from the system. Generally, when using the wet RFG option, the produced combustion gas is extracted before the condensing process, whereas the dry RFG is extracted downstream from the condenser, and both of these options are carried out after the removal of particulates. Under the same volumetric flue gas flow rate, the dry RFG has higher adiabatic flame temperature (AFT) than that in the wet RFG. However, the combination of wet and dry RFG options may provide an attractive option, especially for the PC oxy-fuel power plants. The dry RFG can be utilized to transport coal particles from the mill to the furnace and other miscellaneous uses, while the wet RFG is used to control the combustion temperature due to its high content of water vapour which is much more than in dry RFG. This combined attractive option has recently been used by the Vattenfall project in Germany on the 30 MW oxy-lignite pilot-scale utility boiler [50].

Sturgeon et al. (2009) [33] demonstrated the effect of the RFG ratio on the amount of furnace exit carbon in ash (CIA), on the adiabatic flame temperature, and on the coal residence time in the combustion zone. By increasing the RFG ratio, the flame temperatures and residence times of PC were gradually decreased, resulting in an increase in the exit of CIA from the furnace. In order to avoid making major modifications to the heat exchange equipment of the conventional large-scale boiler, the recycle ratio of flue gas, in the oxy-fuel scenario, has to be precisely determined. The best value of RFG can bring another benefit to the oxy-fuel power plant through eliminating the slagging and fouling formation problems on the water wall and on the convective tubes bank of the boiler such as superheaters and reheaters. This problem can be avoided by keeping the furnace exit gas temperature below the ash melting temperature of coal combustion [10, 12, 16]. This specific temperature is substantially dependent on the ash depositing behaviour of the coal used, as will be seen in the discussion sections of chapters five and seven.

5. CFD modelling investigations on the lab–scale and large–scale furnaces under air–fired and oxy–fuel combustion conditions

In this section, an overview of the research program [10, 11, 51, 52] will be briefly discussed. This research program can be classified to have two main objectives:

• The 3 D numerical simulations of pulverized dry lignite in a 100 kW test facility were conducted to simulate four different combustion environments (air-fired, OF25, OF27, and OF29) and to investigate the temperature distribution levels, species concentrations, and velocity. The commercial CFD software was used to model and analyze all the combustion media. Several mathematical models with the appropriate related constants and parameters were employed for lignite coal combustion. The combustion conditions of oxy-fuel combustion cases were satisfied by modifying the following factors: oxygen concentrations in the feed gas and carrier gas, and recycled flue gas rates [10, 51].

• The commercial CFD code was modified to investigate the Victorian brown coal combustion in a 550 MW tangentially-fired boiler under different combustion media. Several mathematical models such as coal devolatilization, char burnout, combustion chemistry, convection and radiation heat transfer processes, carbon in fly-ash, and thermal and fuel nitric oxides models were developed through subroutines and added to the CFD calculations. The available experimental data from the power plant were used to validate the predicted results under air-fired condition; a good agreement was achieved. The oxy-fuel combustion approach adopted in a 100 kW facility unit (Chalmers' furnace) was applied to the present large-scale furnace in three O_2/CO_2 mixture conditions, namely OF25, OF27, and OF29. These models were implemented to investigate the importance of including such models in conjunction with the newly developed oxy-fuel combustion model [52].

5.1. The pulverized dry lignite combustion in the lab–scale furnace

This subsection describes a comprehensive computational fluid dynamics (CFD) modelling study undertaken [10, 51] by integrating the combustion of pulverized dry lignite in several combustion environments. Four different cases were investigated: an air-fired case and three different oxy-fuel combustion environments (25 % vol. O_2 concentration (OF25), 27 % vol. O_2 concentration (OF27), and 29 % vol. O_2 concentration (OF29). The chemical reactions (devolatilization and char burnout), convective and radiative heat transfer, fluid and particle flow fields (homogenous and heterogenous processes), and turbulent models were employed in 3-D hybrid unstructured grid CFD simulations. The available experimental results [53] from a lab-scale 100 kW firing lignite unit (Chalmer's furnace) were selected for the validation of these simulations.

5.1.1. A review of experimental setup of Andersson [53]

Chalmers 100 kW test facility has been designed to burn both gaseous and pulverized fuels. The furnace is a cylindrical refractory lined drop tube unit with dimensions: 80 (cm) inner diameter and 240 (cm) inner height. Three different oxy-fuel combustion cases are used as follows: OF25 (25% vol O_2 and 72% vol CO_2), OF27 (27% vol O_2 and 71% vol CO_2), and OF29 (29% vol O_2 and 69% vol CO_2). The carrier gas in the air-fired case, which is used to inject the coal, is air with a volumetric flow rate of $83.3*10^{-5}$ (m^3 per sec). While the carrier gas in oxy-fuel combustion cases is recycled flue gas (RGF) with a volumetric flow rate of $66.6*10^{-5}$ (m^3 per sec.) that has the same oxygen concentration (30% vol dry) and recycled flue gas (68% vol dry) for all oxy-fuel combustion cases.

Inlet Flow Field Parameters		Combustion Cases			
		Air	OF25	OF27	OF29
Primary Register	Volume Flow Rate (m³/h)	34.87	28.94	26.85	25.11
	Mean Velocity (m/s)	7.966	6.612	6.134	5.737
	Angular Velocity (rad/s)	433.293	359.615	333.645	312.052
Secondary Register	Volume Flow Rate (m³/h)	81.37	67.54	62.65	58.59
	Mean Velocity (m/s)	4.995	4.146	3.845	3.596
	Angular Velocity (rad/s)	41.14	34.152	31.673	29.608

Table 2. The inlet flow field parameters of all combustion cases for primary and secondary registers of the burner

Generally, the volumetric flow rates of feed oxidizer gases (air or dry RFG for oxy-fuel combustion cases) were 30% for primary register and the rest were through secondary register. The volumetric flow rates of air or RFG through the primary and secondary registers of the burner were decreased gradually by 17%, 23%, and 28% in the OF25, OF27, and OF29, respectively with respect to the air-fired case. The oxidizers / fuel stoichiometric ratio (λ) was kept constant (1.18) for all combustion cases. The initial inlet temperature of dry flue gas at the inlet of burner was around 298.15 K for all combustion tests. The inlet flow field parameters at the primary and secondary registers of the burner, for all the combustion cases, are summarized in Table 2. The combustor is initially fired by using gas (propane) as a pilot-fuel to start up the combusting process before switching to coal.

5.1.2. CFD results and discussion

Figure 4 shows the temperature distributions on a vertical plane through the middle of the furnace. The left hand sides of each image show the air-fired case as the reference. The three different oxy-fuel combustion cases are presented on the right hand sides. The main purpose of this figure is to visualize the overall flame temperature in this axisymmetric furnace.

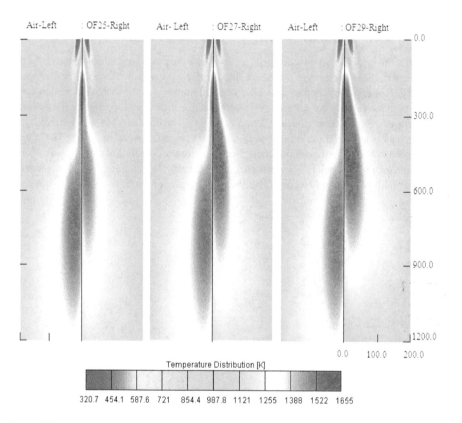

Figure 4. Flame temperature distributions (K) at the vertical cut along the furnace axis for the reference air-fired case (left hand sides) and oxy-fuel combustion environments OF25, OF27, and OF29 (right hand sides), respectively.

The flame started at the burner exit and extended up to the middle of the furnace for the reference air-fired case. The near-burner flame temperature distribution of oxy-fuel combustion (OF25) is closer to that of the air-fired case. Whereas, the near-burner flame temperature of oxy-fuel (OF27) is higher, and the length of the flame is shorter and more confined in the burner exit region. For the OF29 case, the flame temperature distribution is similar to that of OF27 case with marginally higher flame diameter in the near burner region. The peak flame temperature values of the air-fired and OF25 cases were 1603.3 (K), and 1577.5 (K) respectively.

While the peak flame temperature values for OF27 and OF29 were 1666.2 (K) and 1699.1 (K), respectively. These peak flame temperature values are very consistent with the experimental result values [53]. It is clear that when the oxygen concentration in the feed gas stream is increased and the recycled flue gas (RFG) is decreased in the oxy-fuel combustion environments, the flame temperature increases and the shape of the flame is more confined and close to the inlet flow field tip. This phenomenon can be attributed to the higher oxygen content in the near burner reaction zone and higher residence time for the coal due to lower velocity that gives more time to burn. Furthermore, the decreased amount of recycled flue gas (CO_2) will absorb less heat released by combustion, thereby increasing the flame temperature.

In Figure 5, the velocity vectors at the primary and secondary swirl registers, located at the tip of the burner. The differences in the directions of velocity vectors between the primary and secondary registers are related to the values of swirl numbers at the same combustion case. The values of the velocity vectors for both inlet registers are different for different combustion cases and are dependent upon the inlet flow conditions as reported in Table 2. A swirl injection system is widely used in the burning systems in order to increase the mixing and give enough time for oxidizers to burn maximum amount of fuel in that critical zone of the furnace thereby avoid incomplete combustion. The swirl effect, in this study, is certainly used to enhance the turbulent mixing and thereby leads to stabilize the structure of the flame.

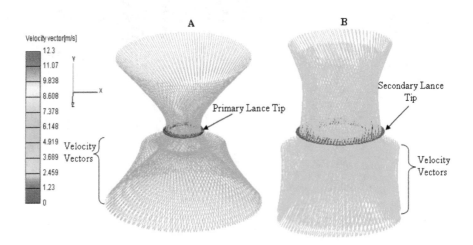

Figure 5. The velocity vectors (m/s) of the primary (A) and secondary (B) swirl registers of the burner in the inlet flow tip of the furnace.

The oxygen concentrations (mass fraction, kg/kg) on a vertical plane through the middle of the furnace are shown in Figure 6. In this figure, the left hand sides of the images are for the reference air-fired case, while the right hand sides are for OF25, OF27, and OF29 respectively. The oxygen concentrations in the air-fired case and OF25 case are approximately similar. These two cases show a delay in the consumption of O_2 in the upper part of the furnace, especially along the centerline of the furnace, compared to OF27 and OF29 cases. The similarity in O_2 consumption between air-fired and oxy-fuel cases is approximately coupled to the flame temperature levels described in Figure 4. In both the OF27 and OF29 cases, the O_2 consumption starts early due to improved ignition conditions and faster combustion leading to the flame to be closer to the burner tip. These results of oxygen concentrations are very similar to that obtained in experiments [53].

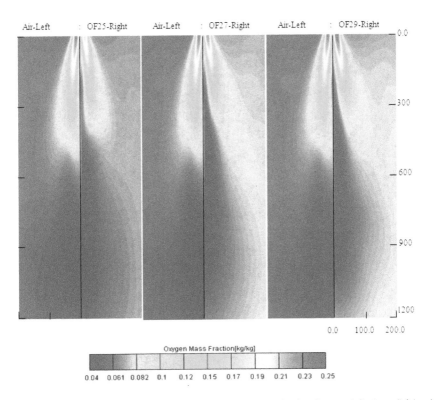

Figure 6. Oxygen mass fraction (kg/kg) in the upper part of the combustor for the reference air-fired case (left hand sides) and oxy-fuel combustion cases OF25, OF27, and OF29 (right hand sides), respectively.

The rapid reduction of oxygen concentration in the near burner region, due to burning of larger amounts of volatile in the OF27 and OF29 cases, negatively affected the oxidation of the residual char in the remaining part of the furnace. Figure 7a shows the char content of coal particles (%) for the air-fired (A) and OF25 (B) combustion cases in the top half of the furnace. Availability of sufficient oxygen led to burn out of residual char by about 900 (mm) into the furnace. Figure 7b demonstrates the residual char for OF27 (C) and OF29 (D) cases. Shortage of the oxygen content led to reduced char burnout, particularly in the OF29 combustion case. The exclusion of the carbon monoxide (CO) as chemical species in the combustion process (i.e. ignoring the Boudouard reaction) of coal, especially in the flame envelop zone (higher temperature region) may have resulted in lower char burnout in the OF27 and OF29 cases. However, these results certainly confirm that the burning out of hydrocarbon gas is faster and low char burnout is achieved at high O_2 concentration of the oxy-fuel combustion cases as reported in the previous findings [17, 19, 24, 54]. As a result, there will be some differences in the total radiation intensities inside the combustor between the reference air-fired case and oxy-fuel combustion case at high O_2 content, especially with high CO_2 concentration in the flue gas.

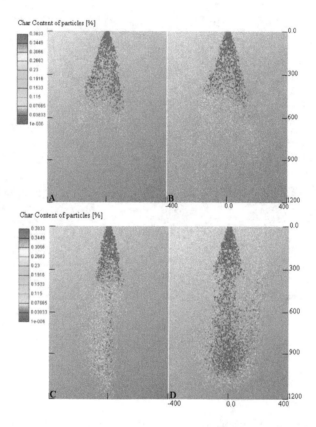

Figure 7. a. Char content of coal particles (%) for the air-fired (A) and OF25 (B) combustion cases over a cross-section of the upper half of the furnace, b. Char content of coal particles (%) for the OF27 (C) and OF29 (D) combustion cases over a cross-section of the upper half of the furnace

The fundamental concept of oxy-fuel combustion technique is mainly to increase carbon dioxide concentration in the flue gas. This technique can be applied, as previously mentioned, by using a mixture of pure oxygen and part of RFG as feed oxidizer gases instead of air to burn with fuel. Therefore, in this study, Figure 8 is clearly showed the increase in CO_2 concentration for all oxy-fuel combustion scenarios (OF25, OF27, and OF29) with respect to the air-fired case. These results were obtained with three-step chemistry mechanisms in terms of homogeneous and heterogeneous coal reactions. As seen, the maximum mass fraction value of CO_2 concentration was about 17.21 % (kg/kg) for air-firing, while for oxy-fuel cases was, in general, about 90.11 % (kg/kg) due to usage of dry flue gas recycled, as implemented in experimental work of Andersson (2007). However, the purity of oxygen (99.5% pure oxygen used in the experiments) and leakage in the fur-nace are relevant parameters to decrease CO_2 concentration, and therefore they should be taken into consideration in design any oxy-fuel combustion boiler. All these CFD results were comprehensively validated against the available experimental data [10, 51].

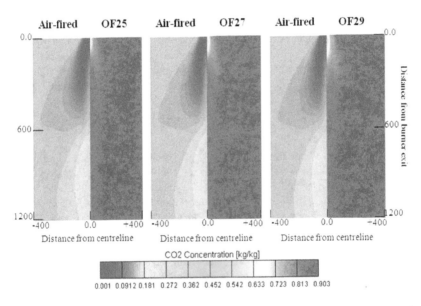

Figure 8. Carbon dioxide concentration (kg/kg) at the vertical cut along the furnace axis for the reference air-fired case (left hand sides) and oxy-fuel combustion environments OF25, OF27, and OF29 (right hand sides) respectively, all dimensions are given in mm.

5.2. The brown coal combustion in a 550 MW tangentially–fired boiler

In order to design such efficient, clean, and economical brown coal combustion systems, the understanding of the brown coal reactivity and behaviour under several combustion condi-tions is required. Generally, brown coal has a number of advantages such as abundance, low-cost, high reactivity, and low sulphur content. In despite of these benefits, a high moisture content (about 60-70 wt %) is the major disadvantage of brown coal. However, in the existing pulverised brown coal (PC) tangentially-fired boiler, a large amount of the hot exit flue gas, typically 50% of the total flue gas generated, is reused to dry the brown coal within the mill-duct system [55- 57]. During that drying process by the hot gas off-takes (HGOTs), a large amount of water vapour is reproduced as well. In order to avoid any flame stability problems inside the combustion chamber, due to that evaporated steam, a fuel-rich mixture (mainly pulverised coal) is passed through the main burner ducts. Whilst a fuel-lean mixture, including water vapour, inert gases, and remaining of PC, is delivered to the inert burner ducts (upper burners). This distribution of the PC and inert gases into the firing system is favourable, particularly in this type of combustion technology. This section focuses on the numerical investigation of a large-scale oxy-fuel furnace. Therefore, the objective of this study is to simulate the brown coal combustion in the large-scale tangentially-fired furnace under several operating conditions. A computational fluid dynamics (CFD) code was used to model and analyse four different combustion environments. A conventional PC combustion and three oxy-fuel combustion scenarios, are known as OF25, OF27, and OF29, were simulated. The validation of the CFD results with the power plant data has been conducted in the air-fired combustion. Four parameters: flue gas composition, gas temperatures, carbon in fly-ash, and HGOT mass flow were compared. Results, for all combustion cases investigated, are compared. The species concentrations, temperature distributions, gas-phase velocity fields, char burnout, NO_x emissions, and radiative heat transfer obtained for all combustion cases were compared.

5.2.1. Boiler description and operating conditions

The tangentially-fired Victorian brown coal 550 MW_e boilers located in the Latrobe Valley mine, Victoria/Australia was used in this simulation study. The geometric description of the CFD model for the boiler, unit 1 at Loy Yang A, is shown in Figure 9. Under maximum continuous rating (MCR) of operating conditions, the unit produces 430 kg/s of steam flow through the main steam piping at 16.8 MPa and 540 °C. The computational domain illustrated in Figure 9 was extended from the furnace hopper up to the top of the tower, passing through the transition of round duct to before the bifurcation at the inlet to the air heaters. In this CFD model, the complex geometric dimensions of the simulated boiler were 98.84 m (height), 17.82 m (width), and 17.82 m (depth), having a net simulated volume of 35,894 m^3 up to the bifurcation point to the air heaters. The tangentially-fired furnace used in this study consists of eight mill-duct systems, two on each side face of the four-sided furnace. For each mill-duct system, there are six separate burners, including three inert burners and three main burners, as well as a hot gas off-take (HGOT) that dries the brown coal. The mill-duct systems were designed for the following purposes: grinding the raw coal into pulverised coal (PC) in the mill, removing the moisture content (62% wt) from the brown coal through the drying shaft, and transporting and distributing the PC. The centrifugal separation system is used to deliver pulverised coal from the grinding mill to the inert and main burners of the furnace. The distribution of PC at both the burner mouths was accompanied by the inert flue gas and water vapour from the drying process in the mill. Approximately 82% of the PC and 34% of the gases is delivered to the main burner (PC burner) and the remaining 18% of pulverised coal and 66% of the gases is transported to the inert burner (vapour burner). This distribution of fuel and gases (fuel-rich mixture) to the main burners is required to maintain combustion stability in the furnace burning Victorian brown coal. Table 3 shows the mass flow distribution of PC and mill gas at each inlet port of the burner ducts. The overall number of vapour and PC burners was 48, while 18 of the total burners were practically out of service, i.e. no fuel is introduced in the latter burners. In the furnace zone, the burners' arrangements on the furnace wall surface were as follows from top to bottom: upper inert burner (UIB), intermediate inert burner (IIB), lower inert burner (LIB), upper main burner (UMB), intermediate main burner (IMB), and lower main burner (LMB).

Burner duct	PC flow (as receive.)		Gas flow	
	Mass flow rate (kg/s)	Distribution ratio (%)	Mass flow rate (kg/s)	Distribution ratio (%)
UIB	0.93	5.7	18.71	22.1
IIB	0.54	3.3	18.22	21.5
LIB	1.57	9.4	18.19	21.5
UMB	4.82	29.7	10.3	12.2
IMB	2.78	17.2	9.34	11.1
LMB	5.62	34.7	9.8	11.6
Total	16.26		84.56	

Table 3. The mass flow rates and the distribution ratios for PC and mill gas at each inlet port of the burner ducts

5.2.2. Cases studies set up

For this numerical study, four different combustion scenarios were selected in order to estimate the performances of the 550 MW$_e$ large-scale boilers under different firing conditions. In the first simulation case, the chemical and physical set up of the boiler operations were completely based on the station data [58]. This first case represents a reference (air-fired) case in investigating the behaviour of brown coal combustion and the boiler performance for the three challenging oxy-fuel combustion scenarios. For the proposed (retrofitted) oxy-fuel combustion scenarios, the thermodynamics set up of the lab-scale oxy-fuel furnace [53], conducted at Chalmers University, was selected in terms of the gas compositions and volumetric flow rates of feed oxidizer gases. The retrofitted oxy-fuel combustion cases were defined as follows: OF25 (25 vol.% O_2 and 75 vol.% CO_2), OF27 (27 vol.% O_2 and 73 vol.% CO_2), and OF29 (29 vol.% O_2 and 71 vol.% CO_2). Detailed information about the Chalmers' furnace and the combustion conditions can also be found in the previous simulation studies [10, 11, 51].

5.2.3. CFD results and discussion

In Figure 10, the distributions of flue gas temperatures are presented along the height of the furnace at the mid cut of X-Z plane for the air-fired, OF25, OF27, and OF29 combustion cases.

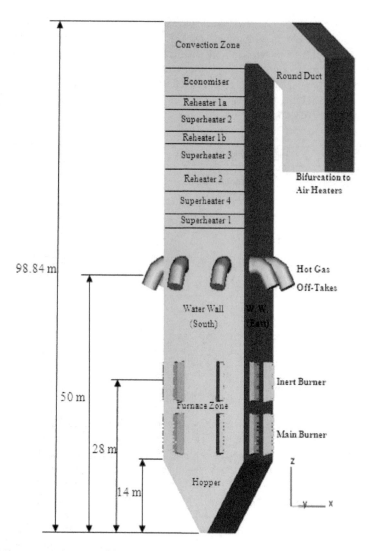

Figure 9. The geometric description of the CFD model for the boiler, unit 1 at Loy Yang A power station

The inlet flow temperatures of gases in the secondary air flows and in the burner gas flows were 473 and 397 K, respectively. Once the reaction processes between PC and oxidizer gases have been started the flame temperature is progressively increased to be at a peak value in the furnace zone as follows: 1864.37 K for air-fired, 1752.0 K for OF25, 1813.3 K for OF27, and 1865.0 K for OF29.

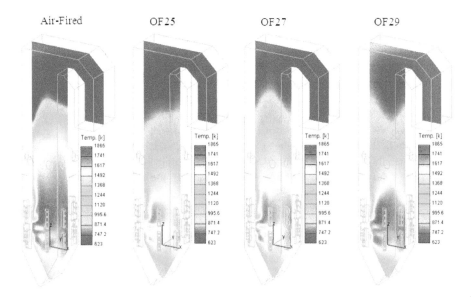

Figure 10. Distributions of the flue gas temperature (K) along the height of the furnace at the mid cut (X-Z plane) for air-fired, OF25, OF27, and OF29 combustion cases

It is clearly seen that a reduction in the levels of the gas temperature occurred when the N_2 is replaced by the CO_2 in the secondary air ducts, particularly in the OF25 and OF27cases examined. That obvious decrease in the gas temperature was mainly due to the higher volumetric heat capacity of CO_2 compared to N_2 in the gas mixture. On the other hand, the maximum gas temperatures between air-fired and OF29 combustion cases were approximately identical. That improvement on the gas temperature of the latter oxy-fuel case was because of increasing O_2 concentration in the feed oxidizer gases. Furthermore, in this study, the oxygen concentration was not only the dominant factor controlling the flame temperature inside the furnace, but also the resident time of coal combustion. However, the inlet flow fields of feed oxidizer gases (O_2/CO_2), in all oxy-fuel cases, were reduced in proportion to the volumetric flow rates by fixed ratios: 83%, 77%, and 72% for OF25, OF27, and OF29, respectively with respect to the conventional firing case. The reduction in the volumetric flow rates in O_2/CO_2 cases has given a sufficient time to burn more coal particles in the combustion zone. However, according to the Chalmers' approach [53], the reduction in the volumetric feed-gas flow rates through the secondary air ducts of the furnace is an efficient method to stabilize and increase the flame temperatures.

Figure 11 shows the distributions of carbon dioxide concentration (kg/kg) along the height of the furnace at the mid cut of the X-Z plane for all cases examined. The recycled flue gas (RFG) (mainly CO_2) used in the oxy-fuel cases has increased the CO_2 concentrations in the flue gases.

The differences in the CO_2 concentrations between the conventional combustion case and the oxy-fuel cases are evident due to adopting the Chalmers' approach in this study. Around five times higher CO_2 is achieved for all oxy-fuel combustion scenarios compared to the air-fired case. This increase of the CO_2 concentrations was also observed in the combined studies of the experimental investigations of Andersson (2007) [53] and numerical modelling of Al-Abbas *et al.* (2011) and 2012 [10, 51] which was conducted on the lab-scale 100 kW firing lignite furnace. The concentrations of CO_2 mass fractions at the furnace exit were equal to 18.84, 85.76, 85.01, and 84.18 wt% for the air-fired, OF25, OF27, and OF29, respectively. Due to the higher capability of carbon dioxide to absorb the combustion heat, this elevated CO_2 in the oxy-fuel cases can potentially increase the protection of the furnace wall against the hot flue gases. However, the heat transfer to the furnace wall for the air-fired and retrofitted OF29 combustion case is technically conformed through two intrinsic aspects: First the gas temperature distributions (in the furnace zone and at furnace exit, see Table 4), and secondly the wall heat flux in difference furnace wall locations.

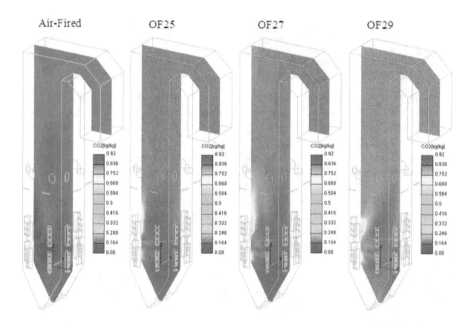

Figure 11. Distributions of carbon dioxide concentration (kg/kg) along the height of the furnace at the mid cut for all cases investigated

Combustion Media	Temp. (K)	CO_2 (wt %)	Net Rad. Heat Flux (MW)	Carbon in Fly-Ash (%)	NO_x (ppm, dry)
Air-Fired (conventional)	635.35	18.84	778.76	2.05	218.58
Oxy-Fuel (OF25)	623.81	85.76	534.3	9.74	96.31
Oxy-Fuel (OF27)	628.27	85.01	596.37	7.04	103.2
Oxy-Fuel (OF29)	631.4	84.18	685.7	5.49	120.0

Table 4. Main predicted CFD results of all combustion cases examined at the final exit plane, and the net radiative heat flux on the furnace wall

6. Conclusion

The oxy-fuel combustion mode can potentially provide significant opportunities for near-zero emissions from the existing and newly-built PC power plants in comparson with the other CO_2 capture technologies such as pre-combustion and post-combustion. In this chapter a computational fluid dynamics (CFD) tool has been developed and validated to numerical turbulent combustion models of gaseous and solid fuels in different furnaces. The numerical investigations of the air-fired and several oxy-fuel combustion environments have been carried out based on the available experimental data. The observations of this research project can provide useful information for improving the combustion characteristics and heat transfer performance of the tangentially-fired boiler under oxy-fuel combustion conditions.

The first objective was to conduct a three dimensional numerical simulation of pulverized dry lignite in a 100 kW test facility in order to provide the confidence to predict the combustion of coal particles in oxy-fuel conditions. The purpose of this study was to simulate four different combustion environments (air-fired, OF25, OF27, and OF29) and to investigate the gas temperature distributions, species concentrations (CO_2, H_2O, and O_2), velocity, and particles consumed in the furnace. The second research objective was to simulate the Victorian brown coal combustion in a 550 MW tangentially-fired boiler under different combustion media. This study focussed on the numerical investigation of a large-scale oxy-fuel furnace. The available power plant data (Staples and Marshall 2010) were used to validate the predicted results under the air-fired condition. The Chalmers' approach in a 100 kW facility unit (Andersson 2007) was selected to the present large-scale furnace in three O_2/CO_2 mixture conditions, referred to as OF25, OF27, and OF29, in terms of aerodynamic conditions and thermodynamic properties.

The findings confirmed that the combustion characteristics and heat transfer processes of oxy-fuel-fired cases can essentially be matched with the conventional combustion characteristics if the appropriate amounts of oxygen and RFG are to be optimised. Compared to the OF25 and OF27 combustion scenarios in the large-scale furnace, the OF29 case showed similar gas temperature levels and radiative heat transfer to that of the air-fired case. This was due to increased residence time of coal particles and O_2 concentrations in the gas mixture. It was also noticed that the higher CO_2 concentrations in the oxy-fuel cases significantly affected the pyrolysis process of coal particles and thus resulted in an increase in the carbon in fly-ash. A remarkable decrease in the NO_x formation was observed because of the elimination of thermal NO process from the oxy-fuel combustion scenarios, as well as low nitrogen content and higher H_2O concentrations in the raw brown coal used. Finally, it can be concluded that the aerodynamic and thermodynamic conditions of OF29 combustion case were favourable, and closely matched the conventional combustion characteristics in several important areas in comparison with the other oxy-fuel-fired conditions.

Acknowledgements

This research program was carried out at Swinburne University of Technology in Melbourne, Australia under the sponsorship of the Iraqi Ministry of Higher Education and Scientific Research (www.mohesr.gov.iq) for developments and innovations in a clean energy sector.

Author details

Audai Hussein Al-Abbas[1] and Jamal Naser[2*]

*Address all correspondence to: jnaser@swin.edu.au

1 Foundation of Technical Education, Al-Musaib Technical College, Babylon, Iraq

2 Faculty of engineering and Industrial Sciences, Swinburne University of Technology, Hawthorn, Victoria, Australia

References

[1] Wall, T.; Liu, Y.; Spero, C.; Elliott, L.; Khare, S.; Rathnam, R.; Zeenathal, F.; Moghtaderi, B.; Buhre, B.; Sheng, C.; Gupta, R.; Yamada, T.; Makino, K.; Yu, J., An overview on oxyfuel coal combustion--State of the art research and technology development. *Chemical Engineering Research and Design* 2009; *87* (8), 1003-1016.

[2] Davison, J., Performance and costs of power plants with capture and storage of CO2. *Energy* 2007; *32* (7), 1163-1176.

[3] Buhre, B. J. P.; Elliott, L. K.; Sheng, C. D.; Gupta, R. P.; Wall, T. F., Oxy-fuel combustion technology for coal-fired power generation. *Progress in Energy and Combustion Science* 2005; *31* (4), 283-307.

[4]   Vitalis, B. In *Overview of oxy-combustion technology for utility coal-fired boilers*, Advances in Materials Technology for Fossil Power Plants - Proceedings from the 5th International Conference, Marco Island, FL, Marco Island, FL, 2008; pp 968-981.

[5]   Amelio, M.; Morrone, P.; Gallucci, F.; Basile, A., Integrated gasification gas combined cycle plant with membrane reactors: Technological and economical analysis. *Energy Conversion and Management* 2007; *48* (10), 2680-2693.

[6]   Ratafia-Brown J; Manfredo L; Hoffmann J; M., R., Major environmental aspects of gasification-based power generation technologies.*SAIC*,2002,*http://www.netl.doe.gov/technologies/coalpower/cctc/cctdp/bibliography/misc/pdfs/igcc/SAIC_Env_Rpt.pdf* (accessed 24 July 2010).

[7]   Kanniche, M.; Gros-Bonnivard, R.; Jaud, P.; Valle-Marcos, J.; Amann, J.-M.; Bouallou, C., Pre-combustion, post-combustion and oxy-combustion in thermal power plant for CO_2 capture. *Applied Thermal Engineering* 2009; *30* (1), 53-62.

[8]   Romeo, L. M.; Bolea, I.; Escosa, J. M., Integration of power plant and amine scrubbing to reduce CO_2 capture costs. *Applied Thermal Engineering* 2008; *28* (8-9), 1039-1046.

[9]   Van Loo, S.; van Elk, E. P.; Versteeg, G. F., The removal of carbon dioxide with activated solutions of methyl-diethanol-amine. *Journal of Petroleum Science and Engineering* 2007;*55* (1-2), 135-145.

[10] Al-Abbas, A. H.; Naser, J.; Dodds, D., CFD modelling of air-fired and oxy-fuel combustion of lignite in a 100 KW furnace. *Fuel* 2011; *90* (5), 1778-1795.

[11] Al-Abbas, A. H.; Naser, J., Numerical Study of One Air-Fired and Two Oxy-Fuel Combustion Cases of Propane in a 100 kW Furnace. *Energy & Fuels* 2012; *26* (2), 952-967.

[12] Nikolopoulos, N.; Nikolopoulos, A.; Karampinis, E.; Grammelis, P.; Kakaras, E., Numerical investigation of the oxy-fuel combustion in large scale boilers adopting the ECO-Scrub technology. *Fuel* 2011; *90* (1), 198-214.

[13] Andersson, K.; Johnsson, F., Process evaluation of an 865 MWe lignite fired O2/CO2 power plant. *Energy Conversion and Management* 2006; *47* (18-19), 3487-3498.

[14] Andersson, K.; Johansson, R.; Johnsson, F.; Leckner, B., Radiation intensity of propane-fired oxy-fuel flames: Implications for soot formation. *Energy and Fuels* 2008; *22* (3), 1535-1541.

[15] Moore, J. D.; Kuo, K. K., Effect of Switching Methane/Oxygen Reactants in a Coaxial Injector on the Stability of Non-Premixed Flames. *Combustion Science and Technology* 2008; *180* (3), 401 - 417.

[16] Hjärtstam, S.; Andersson, K.; Johnsson, F.; Leckner, B., Combustion characteristics of lignite-fired oxy-fuel flames. *Fuel* 2009; *88* (11), 2216-2224.

[17] Zhang, L.; Binner, E.; Qiao, Y.; Li, C.-Z., In situ diagnostics of Victorian brown coal combustion in O2/N2 and O2/CO2 mixtures in drop-tube furnace. *Fuel* 2010; *89* (10), 2703-2712.

[18] Croiset, E.; Thambimuthu, K. V., NOx and SO2 emissions from O2/CO2 recycle coal combustion. *Fuel* 2001; *80* (14), 2117-2121.

[19] Liu, H.; Zailani, R.; Gibbs, B. M., Comparisons of pulverized coal combustion in air and in mixtures of O 2/CO2. *Fuel* 2005; *84* (7-8), 833-840.

[20] Krishnamoorthy, G.; Sami, M.; Orsino, S.; Perera, A.; Shahnam, M.; Huckaby, E. D. In *Radiation modeling in oxy-fuel combustion scenarios*, American Society of Mechanical Engineers, Power Division (Publication) PWR, Albuquerque, NM, Albuquerque, NM, 2009; pp 615-621.

[21] Yin, C.; Rosendahl, L. A.; Kær, S. K., Chemistry and radiation in oxy-fuel combustion: A computational fluid dynamics modeling study. *Fuel* 2011;*90* (7), 2519-2529.

[22] Chui, E. H.; Douglas, M. A.; Tan, Y., Modeling of oxy-fuel combustion for a western Canadian sub-bituminous coal[small star, filled]. *Fuel* 2003; *82* (10), 1201-1210.

[23] Murphy, J. J.; Shaddix, C. R., Combustion kinetics of coal chars in oxygen-enriched environments. *Combustion and Flame* 2006; *144* (4), 710-729.

[24] Tan, Y.; Croiset, E.; Douglas, M. A.; Thambimuthu, K. V., Combustion characteristics of coal in a mixture of oxygen and recycled flue gas. *Fuel* 2006; *85* (4), 507-512.

[25] Cao, H.; Sun, S.; Liu, Y.; Wall, T. F., Computational fluid dynamics modeling of NOx reduction mechanism in oxy-fuel combustion. *Energy and Fuels* 2010; *24* (1), 131-135.

[26] Qiao, Y.; Zhang, L.; Binner, E.; Xu, M.; Li, C.-Z., An investigation of the causes of the difference in coal particle ignition temperature between combustion in air and in O2/ CO2. *Fuel* 2010; *89* (11), 3381-3387.

[27] Zhang, L.; Binner, E.; Qiao, Y.; Li, C. Z., High-speed camera observation of coal combustion in air and O2/CO2mixtures and measurement of burning coal particle velocity. *Energy and Fuels* 2010; *24* (1), 29-37.

[28] Breussin, F.; Lallemant, N.; Weber, R., Computing of oxy-natural gas flames using both a global combustion scheme and a chemical equilibrium procedure. *Combustion Science and Technology* 2000; *160* (1-6), 369-397.

[29] Brink, A.; Kilpinen, P.; Hupa, M.; Kjaeldman, L., Study of alternative descriptions of methane oxidation for CFD modeling of turbulent combustors. *Combustion Science and Technology* 1999; *141* (1), 59-81.

[30] Venuturumilli, R.; Chen, L.-D., Comparison of four-step reduced mechanism and starting mechanism for methane diffusion flames. *Fuel* 2009; *88* (8), 1435-1443.

[31] Bejarano, P. A.; Levendis, Y. A., Single-coal-particle combustion in O2/N2 and O2/CO2 environments. *Combustion and Flame* 2008; *153* (1-2), 270-287.

[32] Li, Q.; Zhao, C.; Chen, X.; Wu, W.; Li, Y., Comparison of pulverized coal combustion in air and in O2/CO2 mixtures by thermo-gravimetric analysis. *Journal of Analytical and Applied Pyrolysis* 2009; *85* (1-2), 521-528.

[33] Sturgeon, D. W.; Cameron, E. D.; Fitzgerald, F. D., Demonstration of an oxyfuel combustion system. *Energy Procedia* 2009; *1* (1), 471-478.

[34] IEA 2009, Oxy-fuel Network, *http://www.co2captureand* (accessed 24 July 2009).

[35] Zhou, W.; Moyeda, D., Process Evaluation of Oxy-fuel Combustion with Flue Gas Recycle in a Conventional Utility Boiler. *Energy & Fuels* 2010; *24* (3), 2162-2169.

[36] Law, C. K., Combustion Physics. Cambridge University Press: 2006.

[37] Arias, B.; Pevida, C.; Rubiera, F.; Pis, J. J., Effect of biomass blending on coal ignition and burnout during oxy-fuel combustion. *Fuel* 2008; *87* (12), 2753-2759.

[38] Molina, A.; Shaddix, C. R., Ignition and devolatilization of pulverized bituminous coal particles during oxygen/carbon dioxide coal combustion. *Proceedings of the Combustion Institute* 2007; *31* (2), 1905-1912.

[39] Rathnam, R. K.; Elliott, L. K.; Wall, T. F.; Liu, Y.; Moghtaderi, B., Differences in reactivity of pulverised coal in air (O2/N2) and oxy-fuel (O2/CO2) conditions. *Fuel Processing Technology* 2009; *90* (6), 797-802.

[40] Kim, G.; Kim, Y.; Joo, Y.-J., Conditional Moment Closure for Modeling Combustion Processes and Structure of Oxy-Natural Gas Flame. *Energy & Fuels* 2009; *23* (9), 4370-4377.

[41] Williams, F. A., A review of flame extinction. *Fire Safety Journal* 1981; *3* (3), 163-175.

[42] Williams, F. A., Progress in knowledge of flamelet structure and extinction. *Progress in Energy and Combustion Science* 2000; *26* (4-6), 657-682.

[43] Hurt, R. H.; Calo, J. M., Semi-global intrinsic kinetics for char combustion modeling. *Combustion and Flame* 2001; *125* (3), 1138-1149.

[44] Hecht, E. S.; Shaddix, C. R.; Molina, A.; Haynes, B. S., Effect of CO2 gasification reaction on oxy-combustion of pulverized coal char. *Proceedings of the Combustion Institute* 2010; *33* (2), 1699-1706.

[45] Andersson, K.; Johansson, R.; Hjärtstam, S.; Johnsson, F.; Leckner, B., Radiation intensity of lignite-fired oxy-fuel flames. *Experimental Thermal and Fluid Science* 2008; *33* (1), 67-76.

[46] Huang, X.; Jiang, X.; Han, X.; Wang, H., Combustion characteristics of fine- and micro-pulverized coal in the mixture of O2/CO2. *Energy and Fuels* 2008; *22* (6), 3756-3762.

[47] Suda, T.; Masuko, K.; Sato, J.; Yamamoto, A.; Okazaki, K., Effect of carbon dioxide on flame propagation of pulverized coal clouds in CO2/O2 combustion. *Fuel* 2007; *86* (12-13), 2008-2015.

[48] Xiumin, J.; Chuguang, Z.; Jianrong, Q.; Jubin, L.; Dechang, L., Combustion characteristics of super fine pulverized coal particles. *Energy and Fuels* 2001; *15* (5), 1100-1102.

[49] Smart, J.; Lu, G.; Yan, Y.; Riley, G., Characterisation of an oxy-coal flame through digital imaging. *Combustion and Flame* 2010; *157* (6), 1132-1139.

[50] Rehfeldt, S.; Kuhr, C.; Schiffer, F.-P.; Weckes, P.; Bergins, C., First test results of Oxy-fuel combustion with Hitachi's DST-burner at Vattenfall's 30 MWth Pilot Plant at Schwarze Pumpe. *Energy Procedia* 2011; *4* (0), 1002-1009.

[51] Al-Abbas, A. H.; Naser, J., Effect of Chemical Reaction Mechanisms and NOx Modeling on Air-Fired and Oxy-Fuel Combustion of Lignite in a 100-kW Furnace. *Energy & Fuels* 2012; *26* (6), 3329-3348.

[52] Al-Abbas, A. H.; Naser, J.; Dodds, D., CFD modelling of air-fired and oxy-fuel combustion in a large-scale furnace at Loy Yang A brown coal power station. *Fuel* 2012; vol. 102, pp. 646-665.

[53] Andersson, K. Characterization of oxy-fuel flames - their composition, temperature and radiation. Thesis, Chalmers University of Technology, Göteborg, 2007.

[54] Chui, E. H.; Majeski, A. J.; Douglas, M. A.; Tan, Y.; Thambimuthu, K. V., Numerical investigation of oxy-coal combustion to evaluate burner and combustor design concepts. *Energy* 2004; *29* (9-10), 1285-1296.

[55] Dodds, D.; Naser, J.; Staples, J.; Black, C.; Marshall, L.; Nightingale, V., Experimental and numerical study of the pulverised-fuel distribution in the mill-duct system of the Loy Yang B lignite fuelled power station. *Powder Technology* 2011; *207* (1-3), 257-269.

[56] Achim, D.; Naser, J.; Morsi, Y. S.; Pascoe, S., Numerical investigation of full scale coal combustion model of tangentially fired boiler with the effect of mill ducting. *Heat and Mass Transfer/Waerme- und Stoffuebertragung* 2009; 1-13.

[57] Ahmed, S.; Naser, J., Numerical investigation to assess the possibility of utilizing a new type of mechanically thermally dewatered (MTE) coal in existing tangentially-fired furnaces. *Heat and Mass Transfer/Waerme- und Stoffuebertragung* 2011; *47*, 457-469.

[58] Staples J.; Marshall L. Herman Resource Laboratories (HRL) Technology: Mulgrave, Victoria Australia, *www.hrl.com.au/*. Report No. HLC/2010/105.

Chapter 3

Application of System Analysis for Thermal Power Plant Heat Rate Improvement

M.N. Lakhoua, M. Harrabi and M. Lakhoua

Additional information is available at the end of the chapter

1. Introduction

In order to improve the performance of a thermal power plant (TPP), it is necessarily to adopt performance monitoring and heat rate improvement. To improve efficiency, the engineer must knew the heat input, the mass of fuel, the fuel analysis and the kW rating generation in order to determine the actual heat rate. After the actual heat rate calculated and understood, losses must be identified and understood. Good communication and teamwork between the engineer and staff within the TPP is essential to success [1-2].

In fact, the heat rate is defined in units of Btu /kWh (KJ / kWh) and is simply the amount of heat input into a system divided by the amount of power generated by of a system [1].

The calculation of the heat rate enable to inform us on the state of the TPP and help the engineer to take out the reasons of the degradation of the TPP heat rate in order to reach the better one recorded at the time of acceptance test when the equipment was new and the TPP was operated at optimum. Therefore, this TPP heat rate value is realistic and attainable for it has been achieved before [1-2].

The global efficiency of a TPP is tributary of a certain number of factors and mainly of the furnace efficiency. Otherwise, there is place to also notice that with regards to the turbine and the alternator that are facilities of big importance in the constitution of a TPP, the degradation of their respective efficiencies hardly takes place long-term of a manner appreciable and this by reason of the ageing of some of their organs as: stationary and mobile aubages usury; increase of the internal flights; usury of alternator insulations, etc.

The objective of this paper is to analyze the different losses of a TPP therefore to implant solutions in order to act in time and to improve its efficiency. Appropriate performance parameters can enable the performance engineer to either immediately correct performance or estimate when it would be cost effective to make corrections. In fact, the performance parameters measure how well a TPP produces electricity.

These actions or decisions are [1]:

• Improve TPP operation;

• Predictive maintenance;

• Comparison of actual to expected performance;

• Improved economic dispatch of TPP;

• Reduce uncertainty in actual costs for better MW sales.

This paper can be loosely divided into five parts. First, we present the functionality of TPP. Second, we present the boiler and steam turbine efficiency calculations. In section 3, we present the methodology of the analysis based on the Objectives Oriented Project Planning (OOPP) method. In section 4, we present the results of the application of system analysis for determining the possible losses for the degradation of the TPP heat rate. The last section presents a conclusion about the advantages and inconveniences of the analysis presented of the TPP heat rate improvement.

2. Functionality of a thermal power plant

Thermal power plant (TPP) is a power plant in which the prime mover is steam driven. Water is heated, turns into steam and spins a steam turbine which drives an electrical generator. After it passes through the turbine, the steam is condensed in a condenser. The greatest variation in the design of TPPs is due to the different fuel sources. Some prefer to use the term energy center because such facilities convert forms of heat energy into electrical energy [3-5].

In TPPs, mechanical power is produced by a heat engine which transforms thermal energy, often from combustion of a fuel, into rotational energy. Most TPPs produce steam, and these are sometimes called steam power plants. TPPs are classified by the type of fuel and the type of prime mover installed (Figure 1).

The electric efficiency of a conventional TPP, considered as saleable energy produced at the plant busbars compared with the heating value of the fuel consumed, is typically 33 to 48% efficient, limited as all heat engines are by the laws of thermodynamics. The rest of the energy must leave the plant in the form of heat.

Since the efficiency of the plant is fundamentally limited by the ratio of the absolute temperatures of the steam at turbine input and output, efficiency improvements require use of higher temperature, and therefore higher pressure, steam.

This overheated steam drags the HP rotor (high pressure) of the turbine in rotation and relaxes to the exit of the HP body of the turbine, so it comes back again in the furnace to be until 540°

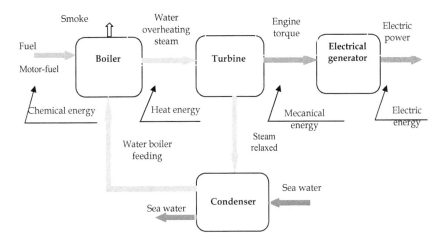

Figure 1. Functionality of a TPP.

after, it will be sent back to the MP body (intermediate pressure) then to the BP body (low pressure) of the turbine.

During these steps, the calorific energy is transformed in available mechanical energy on the turbine. Thus, this mechanical energy will be transmitted to the alternator, being a generator of alternating current, in the goal to produce the electric energy.

After the condensation, water will be transmitted thanks to pumps of extraction in the station of BP to be warmed progressively before being sent back to the furnace through the intermediary of the food pumps.

This warms progressive of water has for goal to increase the output of the furnace and to avoid all thermal constraints on its partitions. And this station of water is composed of a certain number of intersections that is nourished in steam of the three bodies of the turbine. Finally, the cycle reproduces indefinitely since steam and water circulate in a closed circuit.

During this cycle water recovers the calorific energy in the boiler that it restores at the time of its detente in the turbine as a mechanical energy to the rotor of the turbine. The rotor of the turbine being harnessed to the rotor excited of the alternator, the mechanical energy of the turbine is transformed then in electric energy in the alternator.

Turbine constitutes an evolution exploiting principal's advantages of turbo machines: mass power and elevated volume power; improved efficiency by the multiplication of detente floors [6-8].

Indeed, a steam turbine is a thermal motor with external combustion, functioning according to the thermodynamic cycle Clausius-Rankine. This cycle is distinguished by the state change affecting the motor fluid that is the water steam (Figure 2).

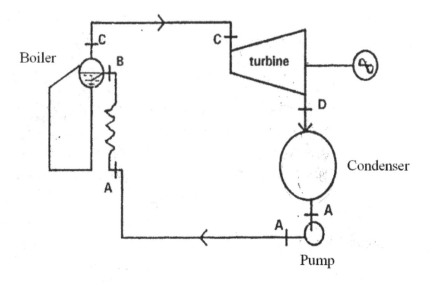

Figure 2. Clausius-Rankine cycle.

The efficiency grows with the steam pressure and with the overheat temperature. However, the increase of these features is limited by the water content in steam in the end of detente.

Indeed, the detente curve can reach the saturation curve with formation of droplets that is harmful to the efficiency of the last floors of detente. The content in liquid water of the mixture must be limited to 15 or 20%. At the end, it is the condenser pressure that fixes the admissible limits of pressure and temperature.

2.1. Boiler accessories

The boiler is a steam generator that assures the spraying of water (Figure 3). At this level operates the transformation of the chemical energy in calorific energy by combustion of a mixture "air-fuel".

Figure 3 presents an example of a boiler in a TPP in Tunisia.

The boiler is composed by different elements:

• The combustion room

It constitutes a surrounding wall of contiguous tubular bundles inside in the water circulates. It is in this combustion room the transformation of the chemical energy in calorific energy by combustion of a mixture "air-fuel". This calorific energy frees a quantity of heat that will be transmitted to water to produce the steam of water in a temperature and under a very determined pressure.

Figure 3. Example of a boiler of a TPP.

- The economiser

It has for role to recover a part of calories remaining in the gases of combustion to increase the temperature of the feeding water what will have for effect the increase of the thermal output of the installation and the elevated thermal constraint suppression in the metal of the reservoir.

- The ball of the boiler

To the exit of the economizer, the water of feeding goes up toward a reservoir situated in the part superior of the boiler called ball of the boiler that constitutes a surrounding wall in sheet metal in which is the liquid phase and the phase steam of the feeding water.

- Superheater / Reheater

It is an intersection of heat constituted of tubular bundles re-serving the gases of combustion directly; therefore submissive to the most elevated temperatures of the combustion room. Steam coming from the ball is humid; it passes therefore in tubes of the heater where its temperature is raised to relatively constant pressure.

After having undergone a first detente in the high pressure body of the turbine, steam comes back to the generator of steam and enter in an intersection called primary reheater of temperature 330°C then it crosses the final re-heater of temperature 540°C, then it is sent toward the intermediate pressure body of the turbine. To the exit of the intermediate pressure body steam passes in the body low pressure of the turbine.

- Burners

The burner is the most important component to light the natural gas fuel-oil. Le role of the burner is of creates a zone of ignition to the sufficient temperature to maintain the combustion and to provide the necessary air mixture. The boiler include seven floors each one is equipped of four burners.

2.2. Steam turbine

A turbine is constituted of a rotor composed of a tree on which is fixed the dawns and a stator of composed of a structural cover of the stationary deflectors, generally organized of two parts according to an axial plan [9-12].

The turbine is composed of a segmented admission tore and a controlled exhaust divergent toward the condenser. The stationary deflector function is to assure all or one of the detente while forming a nozzles network and to modify the direction of the out-flow retiring of the previous floor.

A steam turbine is composed of one or several floors assuring each two functions:

- The steam detente that corresponds to the conversion of the potential energy in kinetic energy;

- The conversion of the kinetic energy in rotation couple of the machine by the mobile aubages.

The steam turbines are often classified in two big categories combined in the same machine:

- Turbines to action in which the detente makes himself solely in the stationary aubages. They are well adapted to strong pressure floors and are better suitable to the debit regulation. Their construction is more expensive and their use for the first floors.

- The jet-propelled turbines in which the detente is distributed between stationary and mobile aubages. The degree of reaction is defined by the distribution of the detente between aubages. They are better suitable to bass pressure floors and their cost is weaker.

The realization of turbines requires the recourse to greatly allied steels (Cr-Ni-Va) to resist the thermal, mechanical constraints (centrifugal force) and chemical (steam corrosion). The first two constraints limit the diameter and therefore the capable debit of the last floors. So dawns besides of one meter of length already put serious problems of realization. Besides, the radial heterogeneity of speeds imposes a variable impact of the dawn that present then a left shape whose machining is complex [9-12].

The principal favour of steam turbines is to be external combustion motors. Of this fact, all fuels (gas, fuel-oil, coal, vestigial, geothermal heat) can be supplied it with steam. The efficiency can reach some elevated values and reduced working expenses (specific consumption of 2300 kcal/kWh and 3400 kcal/kWh for gas turbines). The cost and the complexity of facilities are the most often reserved to the elevated power facilities. But in particular cases, motors and gas turbines are better adapted below about 10 MW.

Figure 4 presents an example of steam turbines in a TPP in Tunisia.

Figure 4. Disposition of steam turbines in a TPP.

3. Boiler and steam turbine efficiency calculations

In this paragraph, we present some methods used in order to determine the boiler efficiency and the steam turbine efficiency [13-15].

3.1. Boiler efficiency calculations

Among the multiple factors that can degrade the boiler efficiency, there is place to mention what follows as an example:

• bad combustion following a bad working of the regulation;

• bad quality of the fuel used;

• heating inadequate of the used fuel;

• flights of water and steam;

• flights of air (comburant);

• encrassement of the boiler;

• encrassement of the air heating device...

Boiler efficiency can be calculated by one of two methods: the Input-Output method or the method of heat losses [1].

3.1.1. Input / output method

The expression of the boiler efficiency of the TPP is given by:

$$R_{ch} = \frac{Output}{Input}. \tag{1}$$

R_{ch}: Boiler efficiency.

The exit is defined by the sum of heats absorbed by the used fluid (water-steam).

The entrance is defined by the total energy introduced in the boiler.

The boiler efficiency is given by the following expression:

$$R_{ch} = \frac{[Q_{eal}(H_{sh}-H_{al}) + Q_{injsh}(H_{sh}-H_{injsh}) + Q_{vrh}(H_{rh}-H_{erh}) + Q_{injrh}(H_{erh}-H_{injrh})]}{F_{h1}.Qf + B_e}. \tag{2}$$

- Q_{eal} : debit water (kg/h);

- Q_{injsh}: debit water injection;

- Q_{injrh}: debit of the water of steam (Kg/h).

- H_{sh}: enthalpy of steam to the exit;

- H_{al}: enthalpy of water in entrance economizer;

- H_{rh}: enthalpy of water (kcal/kg).

- H_{erh}: enthalpy of water exit HP (kcal/kg).

- H_{injrh}: enthalpy of water (kcal/kg).

- F_{h1}: superior calorific power of fuel used (kcal/kg);

- Q_f : debit fued (t/h);

- B_e : total of heat introduced.

3.1.2. Method of heat losses

The boiler efficiency is given by the following expression:

$$R_{ch} = \frac{F_h - C_{per}}{F_h} \qquad (3)$$

F_h : Total energy introduced in the boiler.

C_{per}: Sum of the calorific losses at the level of the boiler.

In the case of fuel, the total heat introduced in the boiler comes from fuel, of the air of combustion and the steam of atomization.

The calorific losses that one meets in a boiler are essentially owed to the heat carried away by the gases of combustion, to the presence of water in fuel as well as the existing humidity in the air of combustion.

3.2. Steam turbine efficiency calculation

The calculation of the efficiency of bodies of the turbine with the difference enthalpy method is very useful for the assessment of the cleaning degree of the steam course in the body of the turbine.

The efficiency of the turbine is defined as the report between the real difference enthalpy (DHIHP) and the isotropy difference enthalpy (DHIHP) of steam crossing the HP body.

For the BP body this method is not applicable because of the title of steam to the BP exit (humid steam).

The efficiency of the HP body (ηHP) is defined as follows [1]:

$$\eta_{HP} = \frac{D_{HRHP}}{D_{HIHP}} \times 100 \qquad (4)$$

The real enthalpy difference in the HP body is given by:

$$D_{HRHP} = H_{vap} - H_{erh} \left(kcal/kg \right) \qquad (5)$$

H_{vap}: enthalpy of steam overheated admission HP turbine.

H_{erh}: enthalpy of steam to overheat HP exit.

The difference isotropy enthalpy in the HP body is given by:

$$D_{HIHP} = H_{vap} - H_{ith} \qquad (6)$$

H_{ith}: final enthalpy of steam overheated for an expansion isotropy (S= constant) of the admission until exit of the HP turbine.

In addition the indicated thermodynamic losses appear in the machine external energy losses provoked mainly by rubbings mechanical landings furniture and flights. The efficiency of the turbine must take into account these losses.

The efficiency of the turbine is:

$$R_t = R_{th} \times R_{vol} \times R_{mec} \tag{7}$$

While the volumetric efficiency (R_{vol}) is equal to:

$$Rvol = 1 - \frac{g}{G} \tag{8}$$

g: debit of flight ; G: debit weight.

4. Methodology of analysis

There are many methods that have been used to enhance participation in Information System (IS) planning and requirements analysis. We review some methods here because we think them to be fairly representative of the general kinds of methods in use. The methods include Delphi, focus groups, SADT (Structured Analysis Design Technique), multiple criteria decision-making (MCDM), total quality management (TQM) and OOPP method (Objectives Oriented Project Planning).

The objective of the Delphi method is to acquire and aggregate knowledge from multiple experts so that participants can find a consensus solution to a problem [16].

A second distinct method is focus groups (or focused group interviews). This method relies on team or group dynamics to generate as many ideas as possible. Focus groups been used for decades by marketing researchers to understand customer product preferences [17].

MCDM views requirements gathering and analysis as a problem requiring individual interviews [18]. Analysts using MCDM focus primarily on analysis of the collected data to reveal users' requirements, rather than on resolving or negotiating ambiguities. The objective is to find an optimal solution for the problem of conflicting values and objectives, where the problem is modelled as a set of quantitative values requiring optimization.

TQM is a way to include the customer in development process, to improve product quality. In a TQM project, data gathering for customers needs, i.e., requirements elicitation may be done with QFD [19].

The SADT method represent attempts to apply the concept of focus groups specifically to information systems planning, eliciting data from groups of stakeholders or organizational teams [20]. They are characterized by their use of predetermined roles for group/team members and the use of graphically structured diagrams. SADT enables capturing of a proposed system's functions and data flows among the functions.

The OOPP method, used in this survey, is considered like a tool of communication, analysis and scheduling of project, whatever is its nature, its situation, its complexity and its sensitivity [21-24].

In this part, we present the OOPP method that we use in order to determine the different losses of the TPP.

4.1. OOPP method

This method is used more and more by several financial backers (World Bank, Union European, bilateral Cooperation…). It is also used to take to terms of development projects, of cooperation (Germany, Canada, Belgium...) or other. It gave a good satisfaction at the time of its exploitation and several researches have been done very well to develop tools and to prove its strength for the scheduling of projects.

The descriptive documentation of the OOPP method, indicate that the logic of the OOPP method is not in principle limited not to a type of a determined problematic. Nevertheless, in practice the method is more appropriated to the following interventions: projects of the technical cooperation and projects of investments with economic and / or social objective.

The OOPP method which is also referred to Logical Framework Approach (LFA) is a structured meeting process. This approach is based on four essential steps: Problem Analysis, Objectives Analysis, Alternatives Analysis and Activities Planning. It seeks to identify the major current problems using cause-effect analysis and search for the best strategy to alleviate these identified problems [21-24].

The first step of "Problem Analysis" seeks to get consensus on the detailed aspects of the problem. The first procedure in problem analysis is brainstorming. All participants are invited to write their problem ideas on small cards. The participants may write as many cards as they wish. The participants group the cards or look for cause-effect relationship between the themes on the cards by arranging the cards to form a problem tree (Figure 5).

In the step of "Objectives Analysis" the problem statements are converted into objective statements and if possible into an objective tree (Figure 6). Just as the problem tree shows cause-effect relationships, the objective tree shows means-end relationships. The means-end relationships show the means by which the project can achieve the desired ends or future desirable conditions.

The objective tree usually shows the large number of possible strategies or means-end links that could contribute to a solution to the problem. Since there will be a limit to the resources that can be applied to the project, it is necessary for the participants to examine these alternatives and select the most promising strategy. This step is called "Alternatives Analysis".

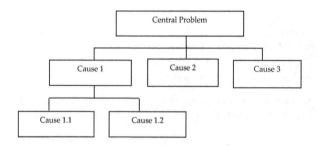

Figure 5. Problem tree of the OOPP method.

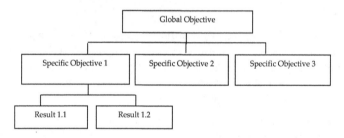

Figure 6. Objective tree of the OOPP method.

After selection of the decision criteria, these are applied in order to select one or more means-end chains to become the set of objectives that will form the project strategy.

After defining the objectives and specifying how they will be measured (Objectively Verifiable Indicators: OVIs) and where and how that information will be found (Means of Verification: MOVs) we get to the detailed planning phase: "Activities Planning". We determine what activities are required to achieve each objective. It is tempting to say; always start at the situation analysis stage, and from there determine who are the stakeholders.

4.2. Some researches on the OOPP method

We present some studies of the OOPP method in IS planning that have been presented in various researches:

Researchers, Gu & al. [25] have presented an object-oriented approach to the development of a generative process planning system. The system consists of three functional modules: object-oriented product model module, object-oriented manufacturing facility model module, and object-oriented process planner.

Researcher, Hill [26] has question the appropriateness of highly structured strategic planning approaches in situations of complexity and change, using the Cambodian-German Health Project as a case study. He has demonstrated the limitations of these planning processes in complex situations of high uncertainty, with little reliable information and a rapidly changing environment.

Researchers, Peffers & al. [27] have used information theory to justify the use of a method to help managers better understand what new Information Technology applications and features will be most valued by users and why and apply this method in a case study involving the development of financial service applications for mobile devices.

Researchers, Killich & al. [28] have presented the experiences and results of the development and implementation of a software-tool for a SME-network in the German automotive supply chain industry. The tool called TeamUp enables the communication of experts as well as the coordination of discussion groups in order to make use of synergetic potentials.

4.3. Refining the OOPP method

The application of the OOPP method to the identification of activities of work stations is important. The management of a system is conditioned notably by ties between its Entities Activities (EA), being able to be according to their hierarchical Specific Objective level (SO), of Results (R), of Activities (TO), of Under-activities (S), of Tasks (T)... These ties are materialized in fact by exchanges of information (If) produced by certain activities and consumed by others. The restraint of these ties requires an extension of the method. This new extension permits to identify the manner to execute these activities and to manage the different phases of the system.

An effort has been provided in order to refine the OOPP method. The OOPP method has been spread and a new denomination MISDIP (Method of Specification, Development and Implementation of Project) was adopted. The MISDIP method adopts the OOPP analysis and the complete it to specify the system of organization, to specify the system of information, and to contribute to its development and implementation [29].

We defined the Method of Informational Analysis by Objectives (MIAO) [30] permitting to elaborate an information matrix that permits to analyze the informational exchange process between activities.

In fact, the identification and the analysis of the information exchanged by the activities indicate the dynamics and the communication between the elements of the system that we propose to study or to manage. So, an information matrix was defined. This matrix establishes a correlation between activities and their information. The information concerning an activity can be classified in two categories:

- The imported information by an activity is supposed to be available: it is either produced by other activity of the system, or coming from outside;

- The produced information by an activity reflects the state of this activity. This last information may be exploited by other activities of the project.

In fact, the information produced by an activity can be considered like a transformation of imported information by this activity.

In order to specify this information, we define an information matrix (Table 1) associated to OOPP analysis enabling to determine the relations between the activities or between the concerned structures, to identify the information sources and to determine the manner in which the information is exploited.

N°	Code	Activity	If$_1$	If$_2$	If$_3$	If$_4$	If$_5$	If$_6$	If$_7$	If$_8$	If$_n$
1		A$_1$	0	0	1	1					
2		A$_2$		0	0		1	0			
3		A$_3$	1	0	0	0		0	0	1	
4		A$_n$									

Table 1. The information matrix of the MIAO.

To make sure of the quality of information system, we define some logic-functional rules reflecting the coherence, the reliability and the comprehensiveness of the analysis by an information matrix in which the rows are relating to activities and the columns to information.

In order to become the exploitation of the information matrix more comfortable, we define the Method of Representation of the Information by Objectives (MRIO) [31] inspiring of the SADT method (Structured Analysis Design Technique) and we define its tools.

5. Results of the system analysis and correctives actions

In this part, we present the results of the system analysis of a TPP whose objective is to determine the possible reasons of the degradation of the TPP heat rate. In fact, all events that are appropriate (preventive or corrective maintenance, exploitation, in conformity of modification, related to working, work stops…) are consigned on GMAO in the TPP in order to constitute the historic and to permit the traceability. This historic has been consulted in the goal to bring a more for the possible problems research.

The objectives tree (presented in linear form) presents ten specific objectives enabling to lead the global objective (GO): TPP heat rate losses identified.

These specific objectives are: Boiler losses identified; Condensate/FW system losses identified; Circulating water system losses identified; Turbine losses identified; Steam conditions losses identified; Electrical auxiliary losses identified; Steam auxiliary losses identified; Fuel handling losses identified; Heat losses identified; Cycle isolation losses identified.

Table 2 presents the OOPP analysis of the TPP heat rate losses.

N°	Code	Activity
1	GO	TPP heat rate losses identified
2	SO1	Boiler losses identified
3	SO2	Condensate/FW system losses identified
4	SO3	Circulating water system losses identified
5	SO4	Turbine losses identified
6	SO5	Steam conditions losses identified
7	SO6	Electrical auxiliary losses identified
8	SO7	Steam auxiliary losses identified
9	SO8	Fuel handling losses identified
10	SO9	Heat losses identified
11	SO10	Cycle isolation losses identified

Table 2. OOPP analysis.

The final production of the application of the OOPP method enabled us to answer clearly to the question « what? ». Then, we presented the different results of the OOPP analysis to enable us identifying the different losses at the level of the TPP and to improve the TPP heat rate.

Figures 7 presents the objectives analysis related to the losses at the level of the boiler.

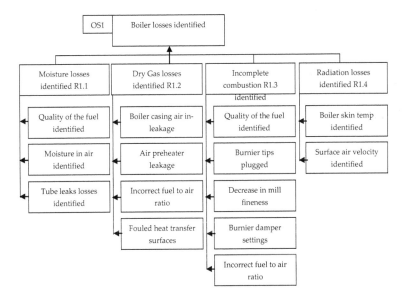

Figure 7. Objective tree of the boiler losses.

For example, the result R1.1 is decomposed in three intermediary results: Quality of the fuel identified; Moisture in air identified; Tube leaks losses identified.

Figures 8 presents the objectives analysis related to the losses at the level of the steam turbine.

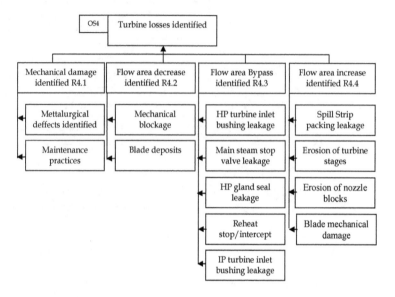

Figure 8. Objective tree of the Turbine losses.

Figures 9 presents the objectives analysis related to the losses at the level of the Electrical auxiliary.

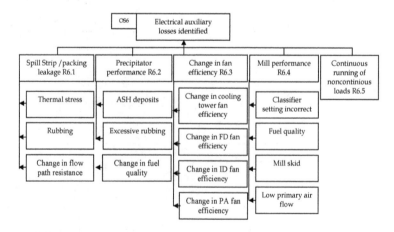

Figure 9. Objective tree of the Electrical auxiliary losses.

Table 3 presents some performance parameters that measure how well the TPP is doing its job in producing electricity. Then, we present in the last column some corrections actions that should be made to improve thermodynamic efficiency and to improve TPP's overall performance.

Parameter	Cause	Test of confirmation	Corrective action
Pressure steam of the admission turbine very high	Instrument error	To verify with the redundant measures To compare with pressures in relation as (pressure exit over heater, ball)	To calibrate instruments
	No consistency with the order point	To verify if floodgates of control of the turbine are opened completely and the pressure remains even high: the coefficient of evaporation debit in the furnace is very low	To verify the order point of the pressure entrance turbine; To see the regulation system
Temperature steam of the admission turbine very low	Instrument error	To verify with the redundant measures To compare with temperatures in relation as the difference with the temperature steam admission turbine must be between 0 and 15°C	To calibrate instruments
	Pressure stem admission turbine high	Pressure steam SH is a performance parameter	
	Debit injection SH	To verify the debit of injection: the debit of injection is a performance parameter	
	Encrassement of exchange surfaces SH	Test integrity of the furnace: factor of tube encrassement	To eliminate the encrassement
	Low excess of air	To verify the excess of air: the excess of air is a performance parameter	
	Order point low	To verify the order point	To increase the value of the order point
Debit injection SH very high	Instrument error	To verify with the redundant measures ; To calculate the debit from the temperature entrance and exit heating	To calibrate instruments
	Order point of steam temperature very low	To verify the order point steam temperature	To increase the value of the order point

Parameter	Cause	Test of confirmation	Corrective action
	Debit injection on manual order point very high	To see system of regulation	To adjust the station of control of injection debit
	Very high air excess	To verify the excess of air: (performance parameter)	
	Low water temperature:		
	a- by-pass HP heating	To verify temperature water before and after by-pass HP	To close floodgate by-pass HP heating device To eliminate the flights
	B High TD heating HP	To verify TD heating HP : (performance parameter)	
	Very low pressure steam admission turbine	To verify pressure steam admission turbine	
	Flight floodgate control of injection	To close the floodgate of insulation ; To verify if the conduct is hot	To repair the floodgate
	Instrument error	To calculate the debit of steam from the temperature of entrance and exit	
	Very low order point of temperature steam	To verify the order point	To increase the value of the order point
Debit injection RH very high	Very high debit injection on manual order point	To verify the control station	To adjust the control station
	Very high air excess	To verify the air excess (performance parameter)	
	Low water temperature:		
	a- by-pass heating HP	To verify temperature water before and after by-pass HP	To close floodgate by-pass HP heating ; To eliminate the flights
	B High TD heating HP	To verify TD heating HP : (performance parameter)	

Parameter	Cause	Test of confirmation	Corrective action
	Very low pressure steam admission turbine	To verify pressure steam admission turbine	
	Flight floodgate control of injection	To close the floodgate of insulation ; To verify if the conduct is hot	To repair the floodgate
	Very low temperature vapeur admission turbine	To verify temperature steam admission turbine	
	Very low efficiency HP turbine	To verify the efficiency of the HP turbine	

Table 3. Diagnosis of performance parameters and corrective actions.

6. Conclusion

The performance of a TPP will begin to decline as the thermal power plant (TPP) begins to age. A good performance program will be able to identify these losses of the degradation of the heat rate. A more accurate knowledge of TPP heat rates can improve economic dispatching costs and ensure that profits are maintained on a daily basis.

In fact, the performance parameters measure how well the TPP is doing its job in producing electricity. Decisions should not necessarily be made only to improve thermodynamic efficiency, but rather to improve TPP's overall performance.

In this paper, we presented an exploration of the ways permitting the improvement of the TPP heat rate. This is why we presented a practical case of a TPP in Tunisia. The objective is to determine the possible causes generating losses and provoking the degradation of the TPP heat rate while using a system analysis method.

To reach this objective, System Analysis methods seem to be a promising way because the major advantage of these kinds of methods is due to the concept of hierarchy activity. These methods permit the complexity of a system to be overcome. In this paper, the application of the OOPP method on a real system, a thermal power plant in Tunisia generates a source of useful information for determining of the possible losses at the level of a TPP. So, research into the application of System Analysis methods must be intensified in order to solve several difficulties and to improve their efficiency.

Author details

M.N. Lakhoua[1*], M. Harrabi[2] and M. Lakhoua[2]

*Address all correspondence to: MohamedNajeh.Lakhoua@ieee.org

1 Université de Tunis El Manar, Ecole Nationale d'Ingénieurs de Tunis, LR11ES20 Analyse, Conception et Commande des Systèmes, Tunis, Tunisie

2 Société Tunisienne de l'Electricité et du Gaz, Tunisie

References

[1] Heat Rate Improvement Reference Manual, EPRI, Palo Alto, CA, TR-109546, July1998.

[2] Heat Rate Improvement Guidelines for Existing Fossil Plants, Electric Power Research Institute, Palo Alto, CA: May 1986. Report CS-4554.

[3] A. Vitaly, Alternative trends in development of thermal power plants, Applied Thermal Engineering, 28, Issues 2-3, 2008, pp. 190-194.

[4] V. Slobodan; P. Nikola and D. Željko; Power Electronics Solution to Dust Emissions from Thermal Power Plants, Serbian journal of electrical engineering, Vol.7. N°2, 2010.

[5] M.N. Lakhoua, SCADA applications in thermal power plants, International Journal of the Physical Sciences, vol.5, N°7, 2010, pp. 1175-1182.

[6] M. T. Khadir and S. Klai, A steam turbine diagnostic maintenance system based on an evolutive domain ontology, International Conference on Machine and Web Intelligence (ICMWI), 2010, pp. 360- 367.

[7] M. Yufeng and L. Yibing, An Hongwen; Statistical analysis of steam turbine faults, International Conference on Mechatronics and Automation (ICMA), 2011, pp. 2413-2417.

[8] H. Qing, D. Dongmei and L. Hong , Research on Web System of Intelligent Diagnosis for Steam Turbine, Chinese Control Conference (CCC2006), 2006, pp. 1271- 1275.

[9] J. Klure-Jensen and R. Hanisch, Integration of steam turbine controls into power plant systems, IEEE Transactions on Energy Conversion, Vol.6, Issue: 1, 1991, pp. 177- 185.

[10] L.N. Bize and J.D. Hurley, Frequency control considerations for modern steam and combustion turbines, IEEE Power Engineering Society 1999 Winter Meeting, Vol.1, 1999, pp. 548- 553.

[11] I.C. Report, Bibliography of Literature on Steam Turbine-Generator Control Systems, IEEE Transactions on Power Apparatus and Systems, Volume: PAS-102, Issue: 9, 1983, pp. 2959- 2970.

[12] W. Jianmei, C. Kai and M. Xinqiang, Optimization Management of Overhaul and Maintenance Process for Steam Turbine, ICIII '08, Vol .3, 2008, pp. 244- 247.

[13] Spécification du calcul de performance ANSALDO Energia 4150 AO VVH I 082.

[14] Norm ASME PTC 4.1- Steam generating units.

[15] FDX 60-000, Normalisation Française, Mai 2002.

[16] R.M. Roth, W.C.I. Wood and A.Delphi, Approach to acquiring knowledge from single and multiple experts, Conference on Trends and Directions in Expert Systems, 1990.

[17] M. Parent, R.B. Gallupe, W.D. Salisbury and J.M. Handelman, Knowledge creation in focus groups: can group technologies help? Information & Management 38 (1), pp. 47-58, 2000.

[18] H.K. Jain, M. R. Tanniru and B. Fazlollahi, MCDM approach for generating and evaluating alternatives in requirement analysis. Information Systems Research 2 (3), pp. 223-239, 1991.

[19] C. Stylianou, R. L. Kumar and M. J. Khouja, A Total Quality Management-based systems development process. The DATA BASE for Advances in Information Systems 28 (3), pp. 59-71, 1997.

[20] K. Schoman, D.T. Ross, Structured analysis for requirements definition, IEEE Transaction on Software Engineering 3 (1), pp. 6-15, 1977.

[21] The Logical Framework Approach (LFA): Handbook for objectives-oriented planning, Norad, Fourth edition, 1999.

[22] ZOPP: An Introduction to the Method, COMIT Berlin, May 1998.

[23] GTZ, Methods and Instruments for Project Planning and Implementation, Eschborn, Germany, 1991.

[24] AGCD, Manuel pour l'application de la «Planification des Interventions Par Objectifs (PIPO)», 2ème Edition, Bruxelles, 1991.

[25] P. Gu and Y. Zhang, OOPPS: an object-oriented process planning system, Computers & Industrial Engineering, Vol.26, Issue 4, pp. 709-731, 1994.

[26] S. H. Peter., Planning and change: a Cambodian public health case study, Social Science & Medicine, 51, pp.1711-1722, 2000.

[27] K. Peffers, Planning for IS applications: a practical, information theoretical method and case study in mobile financial services, Information & Management, Vol.42, Issue 3, pp. 483-501, 2005.

[28] S. Killich and H. Luczak, Support of Interorganizational Cooperation via TeamUp at Internet-Based Tool for Work Groups, Proceedings of the 6th internationally Scientific Conference, Berchtesgaden, May 22-25, Berlin, 2002.

[29] M. Annabi, PIPO étendue : Méthode Intégrée de Spécification, de Développement et d'Implémentation de Projet (MISDIP), International conference on Sciences and Techniques of Automatic control and computer engineering STA'2003, Sousse, 2003.

[30] M.N. Lakhoua, Refining the objectives oriented project planning (OOPP) into method of informational analysis by objectives, International Journal of the Physical Sciences, Vol.6(33), pp. 7550 - 7556, 2011.

[31] Lakhoua, M. N. & Ben Jouida T., Refining the OOPP into Method of Representation of the Information by Objectives, International Transactions on Systems Science and Applications, Vol. 7, No. 3/4, December 2011, pp. 295-303.

Chapter 4

Modernization of Steam Turbine Heat Exchangers Under Operation at Russia Power Plants

A. Yu. Ryabchikov

Additional information is available at the end of the chapter

1. Introduction

The authors' experience in the modernization of commercial heat exchangers in steam turbine units is discussed with the individual features of the operation of these heat exchangers at specific thermal power plants taken into account. The variety of conditions under which heat exchangers operate in specific engineering subsystems requires an individualized approach to the model solutions that have been developed for these systems modernizing. New tube systems for commercial heat exchangers have undergone industrial testing and are in successful operation at a wide range of thermal power plants.

2. Research problem formulation

Extending the service life and increasing the operational efficiency and reliability of heat exchangers in steam turbine units at thermal power plants are of great current interest. Modernizing heat exchangers, operating at thermal power plants, involves less expense than replacing the entire heat exchanger and is mainly aimed at eliminating structural defects that have been identified. This is done both when worn out components are replaced and in the course of planned major overhauls, so that the efficiency and operational reliability of the heat exchangers and the steam turbine unit, as a whole, can be increased substantially.

The major engineering solutions for improvement of commercial heat exchangers in thermal power plants steam turbine units must, first of all, be directed at enhancing the efficiency and reliability of their tube systems, since they are the most subject to wear during operation. The choice of approaches and modern engineering solutions for heat exchangers improvement must meet a number of specifications including the following [1]:

- taking into consideration the specific operating conditions;

- raising the efficiency of heat exchanger while retaining reliability or increasing reliability while maintaining efficiency;

- improving or sustaining repairability and maintainability;

- meeting the standards and rules of the supervisory agency Rostekhnadzor.

The improvement of heat exchangers is a rather complicated system problem. A large number of factors at different levels which influence the efficiency and reliability of a heat exchanger should be taken into account, along with its relationship to the engineering subsystems of steam turbine unit of a thermal power plant. The choice of a modernization scheme involves solving a specific optimization problem where the target function or criterion determining the choice may be any of various characteristics, such as raising the thermal efficiency or the thermal capacity of heat exchanger, or increasing its operational reliability or repairability.

The approach determination for heat exchanger improvement can be divided structurally into three stages: a stage for analyzing the initial data, a stage of computations and accounting for a number of operational and design-engineering factors and a stage of technical and economic analysis and evaluation. One of the most important stages in the analysis is the use of computational methods which take most complete account of the features and behavior of the processes occurring in steam turbine heat exchangers. These methods are used to calculate separately heat transfer coefficients for two heat transfer media, while individual factors affecting the heat transfer are taken into account by introducing corrections to heat transfer coefficients base values. When estimating a choice of improvement method, each specific heat exchanger is considered as a component of a particular engineering subsystem of steam turbine (condensation unit, system for regenerative feed water heating, system of hot water heating, etc.) in which it is employed; here the effect of the proposed modernization scheme on other elements of the subsystem or on the turbine unit as a whole is evaluated.

When improvements are made in heat exchangers at thermal power plants under operating conditions, it is necessary to analyze and take into account a large number of factors and performance indicators, including the specific operating conditions. Most often it is possible to optimize only the basic (thermal and hydraulic) operating characteristics of heat exchanger, leaving the others at levels corresponding to commercial apparatuses.

3. Selection and justification of technical solutions for heat exchanger modernization

Over many years the authors have proposed, tested and brought to realization a series of new, by now commercial, technical solutions for modernizing a large number of different heat exchangers for steam turbine units; these have provided satisfactory solutions of the problems arising during heat exchanger operation, such as [1—3]:

- retaining the external connector dimensions (through pipes and shells) of commercial equipment without change;

- estimation of the choice of material for tube systems and shell components (water chambers);

- use of high efficiency profiled tubes of different materials with optimal profiling parameters (Fig.1);

- optimization of the tube bundles configuration in heat exchangers (in some cases — the increase of heat transfer surface within the shell sizes of commercial heat exchangers);

- optimization of the intermediate partitions arrangement, the rise of tube system vibrational reliability by installing damper belts and clamps;

- employing of a high efficiency method for tubes to tube sheets fixing with the use of annular reliefs made in the tube sheet metal;

- efficient sealing of gaps between the intermediate partitions and the shell (for heat exchangers of "liquid — liquid" type);

- use of special protective coatings.

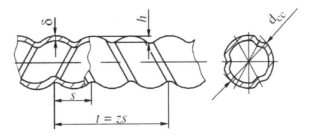

δ is the wall thickness, h is the groove depth, s is the pitch between neighboring grooves, z is the number of profiling threads, t is the thread rolling pitch, and d_{cc} is the circumferential diameter

Figure 1. External view and cross section of profiled twisted tube.

Retrofitting of the oil supply systems of turbine units is intended not only to recover indices characterizing their efficiency and environmental security but also to carry out measures that will improve their characteristics without significant financial and material outlays. These problems may be solved on the basis of technical considerations aimed primarily at oil coolers design.

Our analysis of the efficiency and in-service reliability of over 150 serially produced oil coolers, as well as the data of the Ural All-Russia Thermal Engineering Institute on the vulnerability to damage of oil coolers in service at steam-turbine units, has shown that the elements of these heat-exchangers most susceptible to damage are their tube bundles.

A summary of this data and also the results of tests on a number of apparatuses and experience with their operation under different conditions allowed us to formulate the main lines along which commercially produced oil coolers should be modernized. They are as follows:

- the selection of advanced materials for the elements of the construction;

- the use of heat exchange surfaces that augment heat transfer;

- the creation of optimized layouts for the tube bundles;

- the sealing of gaps in the oil space of the apparatuses;

- the improvement of the reliability of tubes fixing to the tube sheet;

- the protection of the tube sheets against corrosion.

Below, these lines will be considered in more detail.

When selecting the material for the structural elements of an oil cooler, several factors should be taken into account, such as the corrosiveness of the cooling water and the associated corrosion resistance of the heat-transfer tubes; the thermal and hydraulic characteristics of the tubes and their adhesive properties; the compatibility of different materials in one apparatus; the technological features of an assembly of apparatuses with the tubes made of the material selected; and also considerations of cost. The casing of the oil cooler and its parts are usually made of carbon-steel sheets and the tube sheets — of carbon-steel plates or from different types of brass. At most thermal power stations, tubes of brass L68 are installed in the oil coolers; this does not comply with present-day requirements. If a feasibility study is conducted and tubes of corrosion-resistant steels are used in the oil coolers, water chambers and tube sheets made of steels 12Cr18Ni10Ti or Cr23Ni17Mo2Ti can be manufactured. Lately stainless steel tubes are being installed more and more frequently in oil coolers. Considering the higher corrosiveness of the cooling water and the more stringent requirements for environmental safety of thermal power stations, in our opinion, the latter solution is more expedient.

When using stainless-steel tubes in oil coolers, one should bear in mind that the heating capacity of the apparatuses has decreased, because the thermal conductance of the steel is 6 — 7 times lower than that of brass.

Lately, the problem of using titanium alloy tubes in the tube bundles of heat exchangers has been widely discussed. Note that, notwithstanding the high corrosion-resistant and adhesive properties of titanium, several problems in its use still exist, such as protection of the "ferrous" metal of the tube sheets against electrochemical corrosion if it contacts the titanium and the insufficient resistance of titanium to friction wear and its corrosion instability in alkaline solutions with a pH > 10. Note also that the cost of titanium is higher than that of other materials.

The larger overall dimensions of oil coolers for high-power turbine installations required their developers and manufacturers to revise several basic concepts in the design of apparatuses associated, in particular, with the use of new heat exchange surfaces that augment the process of heat transfer, using variously profiled and finned tubes.

We know oil coolers that have tubes with spiral rolled finning, longitudinal welded finning, spiral wire mesh finning, and other kinds of finning. The comparatively new MP-165 and MP-330 oil coolers (for the K-300-240, K-500-240 and K-800-240 turbines manufactured by LMZ), with stainless steel tubes and cross fins manufactured by LMZ instead of the previously manufactured M-240 and M-540 oil coolers, are very efficient products. However, these new designs have, in our opinion, one shortcoming — comparatively low oil velocity, which is due to the non-optimal tube bundle layout in the apparatus.

4. Special features of optimization calculation techniques for heat exchangers

Optimization of the tube bundle layout in heat exchangers, oil coolers included, is one of the most promising ways for improving them. It should be carried out on the basis of a comprehensive calculation of the thermal, hydrodynamic, and reliability characteristics of every given apparatus. The technique for such an optimization calculation has been worked out for oil coolers [4–7] both with plain and with twisted profiled tubes (TPT) that are used in power engineering (see Fig.1). It allows us to account for changes in the parameters of oil in different zones of the apparatus that have been selected downstream of the oil flow. One of the main factors determining the operating efficiency of oil coolers is the leakage of oil through the orifices in the intermediate partition plates and in gaps between the intermediate partition plates and the oil cooler shell. This factor is accounted for in another way.

The permeability of oil through an intermediate partition orifice in the oil cooler was studied at an experimental facility that simulated its geometrical, thermal, and physical performance parameters. The outside diameter of a working tube is 16 mm, the diameter of the orifice in an intermediate partition plate is 16.4 mm, and the thickness of a partition plate is 3 mm. In the experiments we used turbine oil UT30 with a temperature from 35 to 65 °C. The difference in the oil pressures on the partition plate ranged from 850 to 2910 Pa. All the experiments were carried out with the tube centered relative to the orifice in the intermediate partition plate.

Note that the well-known physical laws of liquid flow through an annular slot were confirmed in this study, and the quantitative and qualitative results allowed us to find more exact operating characteristics of oil coolers and to account for them in their designs. The oil flow rate through the annular slot increases as the oil temperature t_{oil} rises and as the difference in pressure across the slot increases. The throughput capacity for the alternative with twisted profiled tubes (TPT) shown in Fig. 1 is considerably higher than that with an annular gap for a plain tube. In this connection, when designing and calculating oil coolers with TPT bundles and plain tubes, we should allow for a sufficiently high level of idle oil leakages through the annular gaps between the tubes and the orifices in the intermediate partition plates.

In contrast to existing methods, the more exact technique developed by us for calculations of oil coolers employs the zone-by-zone approach and accounts for changes in the oil parameters along the oil flow through the tube bundle (Fig. 2). In this case, the oil space of the oil cooler is divided by the partition plates into a number of zones: the input zone (I), the output zone

(N), and intermediate zones (from I to N) in between. The heights of the zones correspond to the distances between the adjacent partition plates.

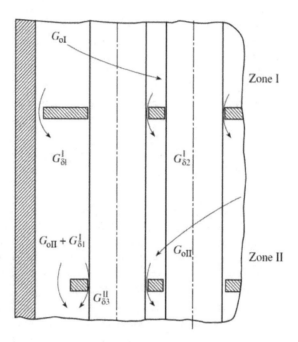

Figure 2. Scheme of oil flow in an element of a tube bundle of an oil cooler with partition plates of the "disc-ring" type and with unsealed gaps. $\delta_1, \delta_2, \delta_3$ are the gaps between the shell and annular partition plate, in the orifices of the annular and disc partition plates; G_o is the oil flow rate through the tube bundle

In every zone, the oil is separated into two or three flows: G_o is the flow through the tube bundle; $G_{oI} = G_o$ and G_{oII} are the flow rates in zones I and II respectively; $G_{\delta 1}{}^I$ is the flow rate in the gap between the annular partition plate and the shell (see zone I in Figure 2); and $G_{\delta 2}{}^I$ and $G_{\delta 3}{}^I$ stand for flow rates in the gaps between the tube and the walls of the orifices in the intermediate partition plates, the annular and the disk ones, respectively. Then the weighted mean temperature of the oil in every flow is calculated. Here, the oil flow rates $G_{oII}, G_{\delta 1}{}^I, G_{\delta 2}{}^I,$ and $G_{\delta 3}{}^I$ were determined by an iterative calculation similar. In the course of the calculation, at a certain viscosity, the average oil flow rate through the tube bundle became negative. This meant that the oil did not reach the extreme tube rows; that is, the apparatuses have zones of stagnation. After making changes in the tube bundle layout, in particular, by decreasing the number of rows of tubes along the depth of the bundle, the thermal hydraulic calculation was repeated.

Calculations executed in accordance with the above technique make it possible to optimize the layout of the tube bundles for the oil coolers and to determine such specifics as the number of rows of tubes that are removed from the center of the tube bundle; the distance between the tubes (the back pitch); the distance between the intermediate partition plates, that is, the number of passes for oil; and others.

5. Methods of heat exchanger reliability improvement

It is known that one of the most important elements determining the operating reliability of an oil cooler is the joint attaching the tubes to the tube sheet. Operating experience with shell-and-tube apparatuses shows that mechanical flaring of tubes in the smooth orifices of the tube sheets (the usual way of attaching the tubes) does not ensure a reliable tightness of the joint. It depends on the following factors: the influence of heat cycles in different directions, the inherent and forced vibrations of the tubes in the bundle and in the apparatus as a whole, the corrosive effect of the cooling water, the natural aging of the material of tubes and tube sheets, etc. [1,2].

An improved tightness and reliability of the joint between the tubes and the tube sheets can be obtained by applying a new technology developed by the specialists of the St. Petersburg State Nautical Technical University. It amounts to flaring the tubes using annular reliefs on the metal in the orifices of the tube sheet, which are formed with the aid of a special tool (Fig. 3).

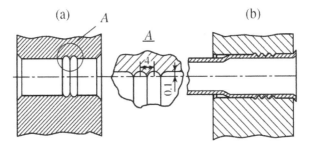

Figure 3. Method for attaching tubes to tube plates. (a) An orifice in the tube sheet with annular reliefs; (b) assembly of attachment of the tube to the tube plate after flaring

To find the optimal dimensions, shape, position, and mechanical characteristics of the annular reliefs, computational studies and experiments were conducted that gave the following results:

- in the area of the annular relief, the contact pressure increases considerably (by two to three times);

- the optimal shape of the relief from the standpoint of most complete imbedding into the tube surface is a triangle with a smooth transition from the base to the walls of the orifice (Fig. 3 a);

- the optimal height of the projection on the relief is in the range from 0.07 to 0.15 mm;

- the maximum increase in contact pressure is obtained when the hardness of the surface of the projection is higher by 30 to 35% on average than the hardness of the metal of the tube sheet;

- the annular relief should be located at a distance of at least 5 mm from the boundary of the zone where the tube is attached (flared) in the tube plate.

We checked the efficiency (reliability) of this technology of attachment on one-tube specimens, pilot modules, and a number of commercial apparatuses having tubes of different materials. On the basis of our investigations, we established that the given method provides a higher tightness than other previously known methods used in oil coolers do, and it is actually not inferior to the combined joint with flaring and welding. Nevertheless, note that the labor required by the proposed method is greater than with usual flaring by 25 to 30%. However, this may be justified.

It is known that the tube sheets of oil coolers can be protected against corrosion due to the cooling water in several ways:

- by rational choice of the material;

- by applying metal coatings and coatings based on paints, glass enamels, different resins, etc.

In oil coolers manufacturing, it is expedient to use corrosion-proof materials for the protection of the tube sheets or to apply metal coatings to them. In operation, the surface may be coated with an epoxy resin or minimum.

6. Discussion of modernization results for power stations heat exchangers

All the above considerations and advanced concepts on improvement of steam-turbine units oil coolers were the basis for modernizing a number of commercially produced oil coolers (MB-63-90; MB-40-60; MP-37; M-21; MOU-12; MO-11; MKh-4), which was done in accordance with our development works. The main measures in retrofitting the oil coolers were taken on the tube bundle, the element most vulnerable to damage in the apparatus. In doing this [1−3]:

- optimization of the layout of the tube bundle is carried out;

- the peripheral annular gap between the intermediate partition plates and the shell of the oil cooler is sealed with an oil-resistant rubber or Teflon plastic (Fig. 4);

- stainless steel twisted profiled tubes (see Fig. 1), whose profile parameters are selected to conform to special apparatuses and conditions of their operation, are employed;

- a new technology for attaching the tubes to the tube plates with annular reliefs is introduced (see above);

- the tube sheets are protected against corrosion.

As an example, in Fig. 5 we find the approximated results of comparative tests of an MP-37 oil cooler for a K-100-90 turbine in the Verkhnetagil district power station before and after retrofitting. From the figure we see that the oil outlet temperature of the retrofitted oil cooler is 0.5 to 2.5°C lower than that of a commercially produced oil cooler. Moreover, note that the operating efficiency of the commercially produced plain-tube apparatus conformed to the designed one. Similar results were obtained for other retrofitted oil coolers, as well.

Up to 10 years of operating experience with modernized oil coolers showed that they ensure the reliable service of the apparatuses without any reduction in their thermal efficiency. No remarks about them were received from the operating personnel.

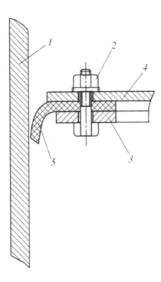

Figure 4. The seal between an annular partition plate and the shell of the oil cooler. (1) shell; (2) bolt with a nut; (3) clamping ring; (4) annular partition plate; (5) oil-resistant rubber (plasticized rubber)

By early 2000, the commercially produced oil coolers used in the lubrication systems of all LMZ K-800-240 turbines operating in the Russian Federation had worked out their service life and the question of replacing them arose. Although commercially produced M-540 oil coolers with wire-loop finning were available, replacement of the old oil coolers with these new ones turned to be inexpedient because the design of M-540 oil coolers had considerable shortcomings. The authors of this paper have developed a highly efficient and environmentally safe MB-270-330 shell-and-tube oil cooler with disk-ring partitions and propose for it to be used to replace commercially produced apparatuses. One specific feature relating to the operation of such heat exchangers with single-phase coolants is that the gaps in the structural members of the inter-tube space have an effect on the coefficients of heat transfer on the oil side.

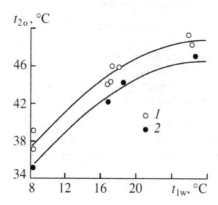

Figure 5. Curves of the oil outlet temperature of the MP-37 oil cooler as a function of the water inlet temperature at the Verkhnetagil district power station. (1) commercially produced oil cooler; (2) modernized oil cooler

Experience gained from modernization of oil coolers for 800-MW turbines has shown that stainless tubes should be used in oil coolers for bringing these apparatuses into compliance with the modern requirements for environmental safety of thermal power stations. In this case, in order to compensate for a lower heat conductivity coefficient of tubes (as compared with that in case of using tubes made of nonferrous alloys), it is recommended that smooth tubes be replaced by twisted profiled tubes (TPTs), which can be made of smooth ones by rolling a helical groove on their external surface. The corresponding protrusions will in this case appear on the inner surface of these tubes (see Fig. 1). The use of TPTs in oil coolers with disk-ring partitions is quite efficient, because it results not only in a smaller pressure drop on the oil side of the apparatus (as compared with its smooth-tube version), but also in a lower temperature of oil at the outlet.

The designs of modernized oil coolers must comply with often contradictory requirements that arise during operation of a thermal power station. This produces a need to carry out calculations of different design versions of oil coolers, which can be performed using zone-to-zone methods. Calculations according to these methods have shown that the number of passes on the oil side has the greatest influence on the thermal- hydraulic characteristics of an oil cooler.

The analysis also shows that there is no sense in sealing the gap in the intermediate partition in oil coolers of "disk-ring" partitions type. The gap between the shell and annular partition, however, should be sealed, because this measure helps improve the efficiency of an oil cooler and obtain lower temperature of oil at the apparatus outlet.

The results of calculations have also shown that the effect of tube material on the oil cooler characteristics is insignificant and that tubes made of Grade 08Cr18Ni10Ti stainless steel are more reliable under specific operating conditions. Changing the diameter of tubes (from 16 to 19 mm) results in that the heat- transfer area becomes 15—18% smaller, and it is hardly advisable. The transverse pitch between the tubes in the bundle is taken equal to 21 mm in the modernized version, which allowed us to obtain the required level of hydraulic resistance on the oil side.

Fig. 6 shows the results obtained from tests of the pilot MB- 270-330 oil coolers at the Perm district power station. It can be seen from the picture that the new oil cooler has better indicators compared with those of the commercially produced M-540 apparatus, which is manifested in a deeper cooling of oil (by 3°C).

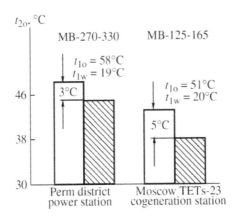

Figure 6. Results obtained from comparative tests of oil coolers. (□) Serially produced oil cooler and (▨) modernized oil cooler; t_{1o} and t_{2o} are the temperatures of oil at the inlet to the oil cooler and at the outlet from it, and t_{1w} is the temperature of water at the oil cooler inlet

Thirteen of the fifteen K-800-240 LMZ turbine units that are installed at Russian thermal power stations now are furnished with MB-270-330 oil coolers. Experience gained from long-term (up to 10 year) operation of the modernized oil coolers has confirmed the advisability of the design and engineering solutions used in these apparatuses that ensure their reliable and efficient operation.

Fig. 6 also shows the results obtained from tests of the MB-125-165 oil cooler that was developed by the authors of this paper and is installed in T-250/300-240 turbines of Ural Turbine Works production instead of the M-240 oil coolers with wire-loop finning that have worked out their service life. The main design and engineering solutions implemented in the MB-270-330 apparatus were used in the course of developing the MB-125-165 oil cooler. It follows from Figure 6 that the oil outlet temperature of the MB-125-165 cooler is 5°C lower than that in the commercially produced M-240 oil cooler. Fifteen MB-125-165 oil coolers have now been manufactured and installed at Russian thermal power stations.

In the following we discuss new designs for other heat exchangers that have been developed and carried out using the above proposed approaches and are well to be recommended for prolonged operation in thermal power plants. The scientific justification for these developments employs a set of theoretical and test stand (physical) studies.

The major problem in modernizing the PSV-315-14-23 hot water heater of KhTZ production K-300-240 turbine at the district heating system of the Reftinskaya State regional electric power plant (GRES) was to develop a design for the preheater which would compensate for the large thermal stresses which develop during startup and shutdown of the preheater while retaining the thermal capacity of the system. Here it was necessary to take into account the fact that during operation under the conditions at the Reftinskaya GRES the main water preheaters essentially operate without being contaminated and they are not cleaned.

This problem is solved by choosing a tube system for the heater (Fig. 7) made of 16/14 mm U-shaped tubes (material CuNi5Fe). The positions of the intermediate partitions are optimized and the bends in the pipes are wound with damping belts in order to reduce the danger of vibrational damage to the tube system. The design of the steam guard shield is changed; this increased the operational reliability of the peripheral sequences of tubes. The casing of the tube system of the heater has vertical and intermediate horizontal baffles whose design facilitates a reduction in the steam return flows alongside the tube bundle, thereby raising its thermal efficiency. The heat exchange surface of the preheater is increased to 335 m^2. A special method is used to fix the tubes to a tube sheet with orifices flared in advance (see Fig. 7).

The major problem in modernizing a VVT-100 heat exchanger intended for distillate cooling in the water cooling system for the generators of the LMZ K-800-240 turbines at the Perm GRES (Fig.8) was to increase the repairability of the heat exchanger while retaining its reliability and thermal capacity. The problem is stated this way because, after disassembly, repair, and reassembly, the gland seal in the VVT-100 heat exchanger is often not leakproof and, with time, the seals on the annular tube partitions are destroyed.

All components of the tube system of the modernized VVT-100M heat exchanger (profiled 16/14.4-mm-diam tubes, tube sheets, special casings) were made of corrosion resistant 08Cr18Ni10Ti (12Cr18Ni10Ti) alloy. The positions of the "disk-and-ring" partitions were determined taking the vibration characteristics of the tubes into account. Special casings which provided additional paths for the cooled distillate in the space between the tubes of the heat exchanger were installed in the upper and lower parts of the tube system to increase the heat transfer efficiency and segmented seals made of teflon tape were installed in the annular partitions to eliminate leakages alongside the tube bundle. A special method with pre-flaring of the orifices was used to fix the tubes in the tube sheets. The upper tube sheet has specially developed gland seals which make it possible to maintain a pressure of up to 2.5 MPa (25 kgf/ cm^2) in tube space. This joint ensures leakproofing of the tube space of the heat exchanger when it is repeatedly assembled and disassembled (see Fig. 8.)

The upper water chamber of the heat exchanger has a removable lid for inspection and cleaning of the tubes without breaking the leaktightness of the tube space. For protection from corrosion the inner surfaces of the water chambers are coated with a special enamel.

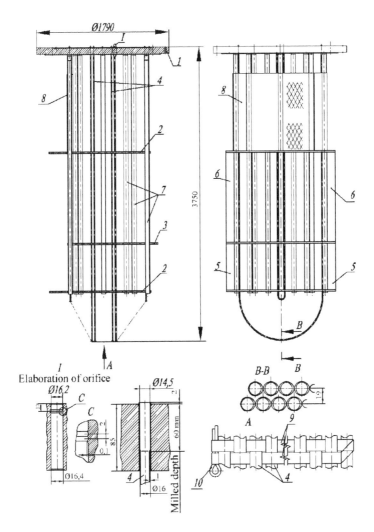

Figure 7. Modernized tube system for a type PSV-315M-14-23 hot water heater: 1 — tube sheet; 2, 3 — disk partitions; 4 — U- shaped heat exchanger tubes; 5, 6 — vertical baffles; 7 — frame tubes; 8 — steam guard shield; 9 — damping belt; 10 — cotter pin

The main task in modernizing the MB-190-250 oil cooler for the lubrication system of the KhTZ K-500-240 turbine at the Reftinskaya GRES (replacing a worn out unit) was to increase its repairability and reliability while retaining its thermal capacity.

The overall and connector sizes for the MB-190M-250 oil cooler (Fig. 9) are the same as those for the commercially manufactured MB-190-250 oil coolers. In the modernized oil cooler the

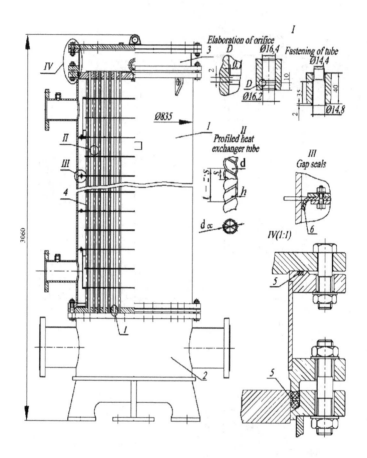

Figure 8. Modernized VVT-100M heat exchanger: 1 — shell; 2 — lower water chamber; 3 — upper water chamber; 4 — tube system; 5 — silicone ring; 6 — teflon seal

major components of the water chambers, tube sheets and 16/14.4-mm-diam heat exchanger tubes are made of corrosion resistant 08Cr18Ni10Ti (12Cr18Ni10Ti) alloy. A special method was used for fixing the tubes to the tube sheets (see Fig. 9.) The system for arranging the "disk-and-ring" partitions and the configuration of the tube bundles were optimized taking the task of modernizing the oil cooler into account. (Plain tubes are installed in the oil cooler, since profiled tubes would lead to an unacceptable increase in the hydraulic resistance on the water side and to a limitation on the flow rate of cooling water in the unit.) In addition, special casings are installed in the upper and lower parts of the tube system to provide additional paths for the cooled oil in the tube space, while segmented seals made of teflon tape are mounted in the annular partitions in order to eliminate leakage along the tube bundle. The oil cooler was made more repairable by changing the design configuration for compensating the temperature expansion of the tube system relative to the shell. For this, the upper tube sheet has an elastic diaphragm (membrane) junction with the shell, while the upper water chamber has a removable cover for inspection and cleaning of the oil cooler tubes without breaking the leaktightness of the oil space. The surfaces of the covers and bottoms that are in contact with the cooling water are coated with a special anticorrosion coating.

Reconstruction of the main ÈP-3-25/75 ejectors coolers of the KhTZ K-300-240 turbine and the ÈP-3-50/150 ejectors coolers of the KhTZ K-500-240 turbine at the Reftinskaya GRÈS was carried out in connection with the end of the coolers' service lives and was related to a need to replace tubes made of CuNi5Fe alloy with tubes made of 08Cr18Ni10Ti stainless steel. The arrangement system for the tube partitions in the steam space of the cooler was optimized taking the change in the tube rigidity into account and the heat exchanger surface was increased as well. When the tubes were flared in the tube sheets a special method was used with the orifices pre-flaring. Profiled heat exchanger tubes were installed in the cooler.

The aim in reconstructing the tube bundles for the fire resistant liquid coolers in the regulatory system for LMZ K-300-240 turbines (Sredneural'sk GRES, Konakovo GRES — horizontal tube bundles) and LMZ K-800-240 turbines (Surgut GRES-2, Nizhnevartovsk GRES, Berezovskii GRES — vertical tube bundles) was to replace tube bundles that had gone past their service lifetimes, as well as to increase their thermal capacity and reliability.

The basic components of the cooler (profiled 16/14.4- mm-diam tubes, tube sheets, special casing) were made of corrosion resistant 08Cr18Ni10Ti (12Cr18Ni10Ti) steel. The configuration of the tube bundle is optimized on the basis of a set of thermal-hydraulic calculations. The thermal capacity of the cooler was raised by increasing its heat exchange surface beyond that of the commercial units. The tubes were flared in the tube sheets using special fixing method with the pre-flared orifices in the tube sheet.

Leakage of the fire resistant liquid off the tube bundle was eliminated by installing a special casing at the annular partitions which ensured that it was joined in a leakproof seal to the annular partitions to prevent the "idle" oil leakage.

Modernization was also carried out for water-to-water heat exchangers of OB-700-1 type deployed in a technological subsystem of Perm state district power station. Heat exchangers of OB-700-1 type are vertical shell-and-tube apparatuses of integral design and this considerably reduces the opportunities of diagnostics, maintenance and repair to be performed by operational personnel of state district power station.

Heat exchanger OB-700-1 is installed at Perm state district power station in the closed system of oil (or fire resistant liquid) cooling for K-800-240 turbines, so for the river Kama protection against oil pollution there is accomplished two-loop system of oil cooling. On passing through the oil coolers the cooling water is directed to two OB-700-1 heat exchangers where it is cooled by circulation water flowing inside the tubes.

The analysis of operation regimes of said heat exchangers showed that during 20 years they were operated with smaller than recommended (nominal) rates of water flow. Because of strong pollution of outside tube surface and significant number of gagged tubes (up to 50%)

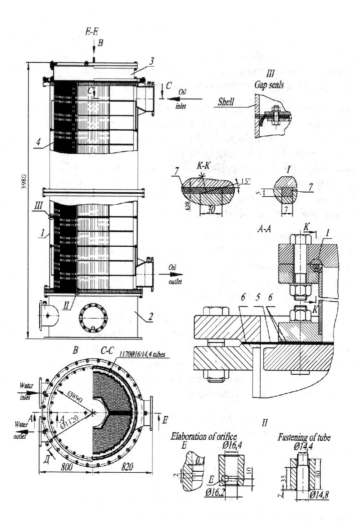

Figure 9. Modernized MB-190M-250 oil cooler: 1 — shell; 2 — lower water chamber; 3 — upper water chamber; 4 — tube system; 5 — diaphragm (membrane); 6 — oil resistant gasket; 7 — silicone ring

the apparatuses have high intertube and inner tube hydraulic resistance. Besides significant corrosion deterioration is observed on tube plates.

According to technical project of Perm state district power station it was necessary when performing OB-700-1 modernization to meet the following basic requirements:

- to reduce surface pollution and lower its affect on thermal efficiency of heat exchangers;

- tube bundle should be disassembling to make it possible to disunite it from the shell provided that all the external connecting dimensions are left without changes;

- to raise the heat exchangers reliability and reduce corrosive affect of the cooling mediums;

- to provide an opportunity of hydraulic test of tube system to be carried out without removing it from the shell;

- to optimize heat exchangers thermal and hydraulic characteristics in order to minimize circulating water consumption.

To meet these challenges during modernization lots of optimizing calculations were carried out with reference to heat exchangers operating conditions at Perm state district power station.

At the concept phase of modernization project the calculations were performed of thermal, hydraulic and reliability characteristics for OB-700-1 heat exchangers. For all the apparatus elements operating under pressure 0.5 MPa durability and reliability characteristics were calculated according to actual normative and technical documentation. On the basis of thermal and hydraulic characteristics optimizing calculations and in order to reduce surface pollution an effective cross section of intertube space was increased due to tube bundle configuration change and to employing of intermediate partitions of «disk-and-ring» type. Also heat transfer surface was reduced from 896 m^2 to 480 m^2, demanded operational characteristics being unchanged. Tube system vibration characteristics calculation permitted to choose a number of its constructive parameters such as thickness of intermediate tube partitions, technological annular clearance between tube and tube sheet orifice, distance between intermediate tube sheets.

According to the design project developed by authors the tube system of modernized heat exchangers OB-700M (Fig. 10) employs intermediate tube partitions of «disk-and-ring» type, welded to frame tubes of tube bundle. Absence of leakage between intermediate tube partitions and cylindrical shell is provided by means of special sealing (Fig. 10).

During modernization of OB-700-1 type heat exchangers for Perm state district power station the following technical conceptions have also been accomplished:

- connecting and overall dimensions of the new (modernized) tube system make possible its installation into the existing cooler shell;

- tubes are made of copper-nickel alloy CuNi5Fe possessing high corrosion resistance. The anticorrosive covering is rendered on tube sheets and on internal surfaces of water chambers to increase their corrosion stability;

- for increasing of tube and tube sheet jointing tightness a special way is used employing reliefs rolled on a tube plate aperture surface;

- inward flange (item 5 in Fig. 10) is used which allows after bottom water chamber removal to carry out hydraulic test without tube system lifting out of the shell;

- for bottom water chamber as well as for inward flange sealing a silicon cord is employed enabling repeated disassembly of the chamber without replacement of sealing material;

- hatches of 500 mm in diameter are made in the top water chamber for tube sheets survey and cleaning.

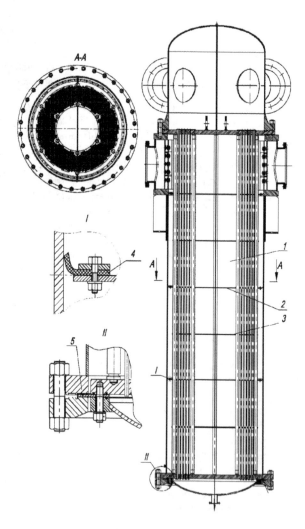

Figure 10. Heat exchanger OB-7OOM. 1 — tube system assembly, 2 — intermediate tube partition of «ring» type, 3 — intermediate tube partition of «disk» type, 4 — special sealing, 5 — inward flange

According to the described concept in 2005 modernization and advanced development were performed of two samples of OB-700M type heat exchangers. After the modernization some starting-up and adjustment tests on OB-700M heat exchangers were carried out at Perm state district power station. For this purpose the heat exchanger was equipped with measuring devices permitting complete thermal balance definition. During the tests the temperatures and flow rates were measured for cooling and cooled water as well. The maximal divergence of heat flow values did not exceed 9.2%.

Test results showed that modernized heat exchangers are capable to function effectively, i.e. to cool the water to temperature demanded by technological process.

The measured values of heat exchanging mediums temperatures are close to designed ones. Thus, for example, the difference between designed and measured water temperature does not exceed 0.6°C.

7. Conclusion

The article presents the results of selection and justification of a number of technical solutions for power station heat exchanger modernization. The calculation technique is laid down for oil coolers with TPTs. The results are shown of preproduction tests for the new heat exchangers as well as a number of advanced and refined designs of oil coolers and water heaters.

The heat exchangers of steam turbines modernized according to described concepts have undergone industrial tests and are in successful use at various thermal electric power plants of Russia. Prolonged (up to 20 years) operating experience for more than 500 developed and manufactured apparatuses has confirmed the validity of the technical solutions, as well as high efficiency and operational reliability of these heat exchangers.

Nomenclature

TPT — twisted profile tube

δ — wall thickness

h — depth

s — pitch between the neighboring grooves

z — number of profiling threads

t — thread rolling pitch

d_{cc} — circumferential diameter

$\delta_1, \delta_2, \delta_3$ — the gaps between the shell and annular partition plate, in the orifices of the annular and disc partition plates correspondingly

G_o — oil flow rate through the tube bundle

t_{1o} and t_{2o} — oil temperatures at the inlet and outlet of the oil cooler

t_{1w} — temperature of water at the oil cooler inlet

Author details

A. Yu. Ryabchikov

Address all correspondence to: lta_ugtu@mail.ru

"Turbines and Engines" department, Urals Federal University (UrFU), Yekaterinburg, Russia

References

[1] Ryabchikov, A.Yu., Aronson, K.E., Brodov, Yu.M., Khaet, S.I., Blinkov, S.N., Zhelon-kin, N.V. "Modernization of Heat Exchangers in Steam Turbine Units Taking Features of Their Operation at Specific Thermal Power Plants into Account", 2010. Thermal Power Stations. № 3. p.p. 28—32.

[2] Brodov, Yu.M., Aronson, K.E., Ryabchikov, A.Yu., Plotnikov. P.N., Bukhman, G.D. "Retrofitting of Oil Coolers for Steam-Turbine Installations". 1998. Thermal Engineering (English translation of Teploenergetika). № 12. p.p. 1012—1016.

[3] Aronson, K.E., Brodov, Yu.M., Ryabchikov, A.Yu., Brezgin, V.I., Brezgin, D.V. "Experience Gained from Development of Modernized Oil Coolers for the Oil Supply Systems Used in 800-MW Turbines", 2009. Thermal Engineering (English translation of Teploenergetika). № 8. p.p. 636—643.

[4] Brodov, Yu.M., Gal'perin, L.G., Savel'ev, R.Z., Ryabchikov, A.Yu., "Hydrodynamics and Heat Transfer in Film Condensation of Steam on Vertical Twisted Tubes," 1987. Thermal Engineering (English translation of Teploenergetika), 34 (7), pp. 382—385.

[5] Berg, B.V., Aronson, K.E., Brodov, Yu.M., Ryabchikov, A.Yu. "Condensation of Steam in Transverse Flow to a Vertical Tube," 1988. Heat transfer. Soviet research, 20 (4), pp. 472—479.

[6] Brodov, Yu.M., Aronson, K.E., Ryabchikov, A.Yu., "Heat Transfer with a Cross Flow of Steam Condensing on Vertical Tubes". 1989. Thermal Engineering (English translation of Teploenergetika), 36 (5), pp. 270—273.

[7] Brodov, Yu.M., Aronson, K.E., Ryabchikov, A.Yu., Lokalov, G.A. "An experimental investigation of heat transfer for viscous liquid flowing over bundles of smooth and

profiled tubes as applied to turbine unit oil coolers", 2008. Thermal Engineering (English translation of Teploenergetika), 55 (3), pp. 196−200.

Sustainable Power Generation and Environmental Aspects

The Effect of Different Parameters on the Efficiency of the Catalytic Reduction of Dissolved Oxygen

Adnan Moradian, Farid Delijani and
Fateme Ekhtiary Koshky

Additional information is available at the end of the chapter

1. Introduction

Many equipment, in particular steam generators of power reactors and water cooled stator windings in turbine generators, suffer from various forms of corrosion induced by the presence of dissolved Oxygen [1]. Removal of dissolved Oxygen (DO) from water is a necessary process in many industries including pharmaceutical, food, power and semiconductor. Acceptable levels of DO vary depending on the intended use of the water; in the power industry, for example, removal of DO is necessary to prevent corrosion in boilers and pipes, and levels of around 5 ppm are necessary. In comparison, ultrapure water, as used in the washing of silicon wafers in the semiconductor industry, is perhaps the most demanding in terms of DO level with some applications requiring extremely low DO levels of around 0.1 ppb [2, 3].

Dissolved Oxygen can be removed from water using a variety of methods broadly grouped into chemical, physical and hybrid systems, which make use of a combination of these methods. Physical methods used including thermal degassing, vacuum degassing or nitrogen bubble deaerations are traditionally carried out in packed towers. Disadvantages with these methods include high operating costs and a small surface area per unit volume. Using these physical methods, it is difficult to reduce the DO concentration from mg/L to µg/L levels. Physical methods have inherent deficiencies of being bulky, costly and inflexible in operation [4]. Recently, hollow fiber membrane contactor with high efficiency and some other advantages have been utilized to remove dissolved Oxygen but their use is still not common [5]. Chemical methods such as the use of sodium sulphite hydrazine, carbohydrazide, β-ketogluconate, and gallic acid or catalytic reduction offer a significant disadvantage in that a further, often toxic, impurity is introduced into the system [6, 7].

Sodium sulfite is another agent for Oxygen removal available in the industry which accounts for its use in low pressure systems. Using Sodium Sulfite at high pressure faces two problems. At first, the consumption of this agent increase solids in the circulating boiler system where the controlling of this parameter at proper range is more important. Second, at high pressure boilers, Sodium sulfite breaks down to form Sulfur dioxide or Hydrogen Sulfide those both are corrosive gases which leave the boiler with steam resulting in low pH steam and condensate and cause the corrosion throughout the system.

Hydrazine (N_2H_4) is a powerful reducing agent that reacts with dissolved Oxygen to form nitrogen and water as follows:

$$N_2H_4 + O_2 \Rightarrow N_2 + 2H_2O$$

At high temperature and pressure, ammonia is also formed, which increases the feedwater pH level, reducing the risk of acidic corrosion. Hydrazine also reacts with soft haematite layers on the boiler tubes and forms a hard magnetite layer, which subsequently protects the boiler tubes from further corrosion. This occurs as a result of the chemical reaction:

$$N_2H_4 + 6Fe_2O_3 \Rightarrow 4Fe_3O_4 + N_2 + 2H_2O$$

Thus in order to reduce or remove the Ammonia, the injection of hydrazine should be reduced or stopped. Recently use of Hydrogen in the presence of catalyst is an attractive method; catalytically recombining the dissolved Oxygen with Hydrogen to form water is an attractive method, as it produces no byproduct [8,9]. Also, the catalytic method can reduce Oxygen levels below one part per billion [10]. In this study removal of dissolved Oxygen in water through reduction catalytic method, was investigated. Also the operation condition such as temperature, pressure, flow rate of water, contact time and inlet dissolved Oxygen concentration was studied on the efficiency of the catalytic reduction of dissolved Oxygen. The part of the results of this work has been published in International Journal of MEJSR [11].

2. Catalytic reduction of dissolved oxygen

In the process of catalytic reduction of dissolved Oxygen, Hydrogen and Oxygen react in the presence of a catalyst to produce water:

$$O_2 + 2H_2 \rightarrow 2H_2O \qquad \Delta H = -68.3 kcal/mol$$

The production of reaction is water that has no adverse effect on the system. An experimental setup was constructed to investigate the effective parameters in the catalytic reduction of dissolved Oxygen. Figure (1) shows the overview of this process. Water is pumped from the water tank to the Hydrogen/water mixer. The mixer is a concurrent gas–liquid upflow packed bed. The water saturated both with Oxygen and Hydrogen then enters the catalytic resin vessel where the Hydrogen and Oxygen react in presence of the catalyst. The reaction is catalyzed using 1.5 liter of K6333 resin catalyst (Lanxess Co. [12]). All results have been obtained under the following conditions:

- Operating pressure: 2- 3 bar

- For determination of remaining Oxygen in the product, used of ASTM D888-81

- Hydrogen gas with 99.99% purity

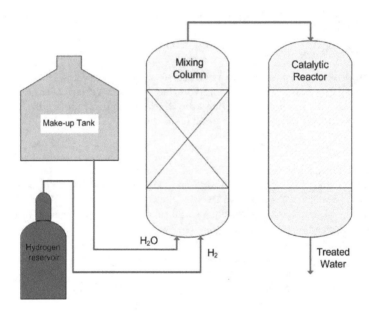

Figure 1. Overview catalytic process to remove dissolved Oxygen

3. Results and discussion

3.1. Effect of temperature and pressure

In order to achieve optimum operation condition, various experimental is implemented at different pressures and temperatures. As shown in Figure (2) the proper temperature range for this system is between 10-50 °C. At temperature lower than 10 °C, the efficiency and the absorption of Oxygen would be reduced. This is due to the fact that at low temperature, the reaction between Oxygen and Hydrogen is no more possible and the catalyst is not able to accelerate this reaction. In other hand further increase in temperature upper than 50 °C may cause damage in the catalyst.

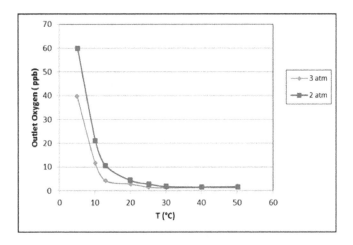

Figure 2. The effect of temperature on outlet Oxygen

Also as can be seen in Figure (3), the appropriate pressure range for this process is between 1.7-3 atm. As previously described the solution of Hydrogen in water is a function of the pressure based on Henry's law. The Henry constant for Hydrogen solution is (7.07 ×10^4 atm.mol H_2O/molH_2 (25°C)), therefore the solution of this gas in water take places slowly, Indicating that the minimum pressure required to maintain the process in a liquid phase. These results also show that at upper temperatures, the pressure should be higher than that at low temperatures. Anyway the increase of Hydrogen solution with pressure results in the system efficiency improvement.

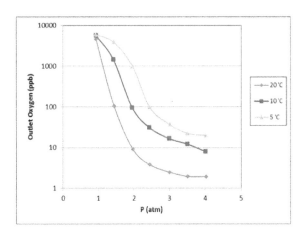

Figure 3. The effect of pressure on outlet Oxygen

3.2. Effect of water flow rate

Other experiments are implemented to illustrate the effect of water flow rate. At first, the inlet Oxygen level is reduced while other parameters remain constant. The results shown in Table (1) indicate that in this case the remaining Oxygen is not upper than the limited level when the water flow rate is 120 lit/hr. In other experiment the water flow rate would be 240 lit/hr and other parameters are constant. In this condition as can be seen in Table (2), it is possible to reduce dissolved Oxygen by means of increasing the water flow rate at low inlet Oxygen level. In this condition the process would be more efficient.

Sample No.	Sampling Time (min)	Inlet Dissolved Oxygen (ppm)	Temperature (°C)		Pressure (bar)		Hydrogen Injection Pressure (bar)	Water flow rate (lit/hr)	Remaining dissolved Oxygen (ppb)
			Mixing Column	Catalytic Reactor	Mixing Column	Catalytic Reactor			
1	35	7.2	22	23	1.5	1.5	2	120	8
2	40	6.9	22	23	1.5	1.5	2	120	6
3	45	6.2	22	23	1.5	1.5	2	120	5
4	50	5.8	22	23	1.5	1.5	2	120	4
5	55	5.2	22	23	1.5	1.5	2	120	4
6	60	4.6	22	23	1.5	1.5	2	120	4
7	65	4.1	22	24	1.5	1.5	2	120	3
8	80	3.5	22	24	1.5	1.5	2	120	2

Table 1. The performance of system at various values of inlet Oxygen

Sample No.	Sampling Time (min)	Inlet Dissolved Oxygen (ppm)	Temperature (°C)		Pressure (bar)		Hydrogen Injection Pressure (bar)	Water flow rate (lit/hr)	Remaining dissolved Oxygen (ppb)
			Mixing Column	Catalytic Reactor	Mixing Column	Catalytic Reactor			
1	35	7.2	22	22	1.5	1.5	2	240	28"/>
2	40	6.8	23	23	1.5	1.5	2	240	28"/>
3	45	6.1	23	23	1.5	1.5	2	240	28"/>
4	50	5.8	23	23	1.5	1.5	2	240	28"/>
5	55	5.1	23	23	1.5	1.5	2	240	24"/>
6	60	4.5	24	23	1.5	1.5	2	240	22
7	65	4	24	24	1.5	1.5	2	240	16
8	70	3.5	25	24	1.5	1.5	2	240	10

Table 2. The effect of increasing of water flow rate

Also the performance of system is investigated at higher level of inlet Oxygen by providing other condition for the results of Table (2). For example when the inlet Oxygen has value of 4.5 ppm, the remaining Oxygen level is upper than limited level (22 ppb). In this case the level of inlet Oxygen and the flow rate of water are maintained constant and other condition have been effectively changed. As shown in Table (3), the performance of system has been improved with increasing both the pressure and temperature, so the remaining Oxygen level reaches lower than 20 ppb but these changes occurred gradually and slowly. This could be explained by the fact that the pressure increasing is effective only to a certain value and further increase after that only cause the further solution of Hydrogen without any effect on the system performance. In other hand, although the increase of temperature is effective for reaction of Hydrogen and Oxygen but the increase of temperature higher than 50 has an inverse effect on Hydrogen solution in water and any increase of temperature at constant pressure results in reduction of Hydrogen solution.

Sample No.	Sampling Time (min)	Inlet Dissolved Oxygen (ppm)	Temperature (°C)		Pressure (bar)		Hydrogen Injection Pressure (bar)	Water flow rate (lit/hr)	Remaining dissolved Oxygen (ppb)
			Mixing Column	Catalytic Reactor	Mixing Column	Catalytic Reactor			
1	40	4.5	24	24	1.5	1.5	2	240	26
2	45	4.6	26	26	1.6	1.6	2	240	22
3	50	4.5	28	28	1.7	1.7	2	240	20
4	55	4.6	30	30	1.8	1.8	2.1	240	18
5	60	4.5	32	32	1.8	1.8	2.2	240	17
6	65	4.5	34	34	2	2	2.3	240	15
7	70	4.6	36	36	2.1	2.1	2.4	240	13
8	75	4.5	40	40	2.2	2.2	2.5	240	12

Table 3. The effect of other parameters on system performance (inlet Oxygen level: 4.5 ppm)

3.3. The system performance in low level of inlet oxygen and high water flow rate

In order to investigate the system performance in low level of inlet Oxygen and high water flow rates, another experiment is implemented with the value of 1ppm of inlet Oxygen and the water flow rate of 640 lit/hr. The results listed in Table (4) show the reduction of dissolved Oxygen with time. However the residence time is reduced in both Hydrogen and catalytic towers because of increasing in water flow rate. Thus under this condition, the time is not enough for complete solution of Hydrogen and reaction with Oxygen. For example after 75 min, the remaining Oxygen is 22 ppb while the sufficient time for this reduction has obtained almost 30 min before.

Sample No.	Sampling Time (min)	Inlet Dissolved Oxygen (ppm)	Temperature (°C)		Pressure (bar)		Hydrogen Injection Pressure (bar)	Water flow rate (lit/hr)	Remaining dissolved Oxygen (ppb)
			Mixing Column	Catalytic Reactor	Mixing Column	Catalytic Reactor			
1	40	1	24	24	1	1	2	640	>28
2	45	1.1	24	24	1.1	1.1	2	640	>28
3	50	1	25	25	1.2	1.2	2	640	>28
4	55	1.1	25	25	1.3	1.3	2	640	>28
5	60	1	26	26	1.4	1.4	2	640	>28
6	70	1	27	27	1.5	1.5	2	640	28
7	80	1	27	27	1.6	1.6	2	640	26
8	90	1	28	28	1.7	1.7	2	640	22

Table 4. The system performance in low level of inlet Oxygen and high water flow rate

3.4. Effect of contact time

An important parameter to control the process is the contact time that influenced by column dimensions, volume of resin and the linear velocity of water. In fact the contact time is fluid residence time in the catalyst column in which DO on the catalyst surface reacts with Hydrogen. The results of experiments in different conditions are shown in Figure (4). In all conditions efficiency is the Oxygen absorption relative to the input of dissolved Oxygen in water.

As seen in Figure 4 in the contact time of less than 45 seconds, the decrease in dissolved Oxygen levels below 5 ppb is not possible. It is hard to reduce the dissolved Oxygen level of less than 10 ppb when the exposure time is less than 30 seconds. Therefore, the ideal time should be considered based on desired reduction of dissolved Oxygen. For use in steam power plants that dissolved Oxygen levels should be less than 5 ppb, time is at least about 45 seconds.

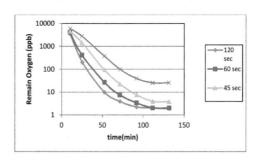

Figure 4. The effect of contact time on remaining Oxygen

3.5. Effect of inlet dissolved Oxygen concentration

The amount of dissolved Oxygen in the water inlet at the start time of the system is effective on the operating time. In other words, the time required to reach the limit concentration for the remaining Oxygen in the outlet water of the catalytic column, is influenced by the dissolved Oxygen in water input. In a fixed volume of catalyst, the less amount of dissolved Oxygen in water, a higher amount of the water flows through the catalyst bed. In this condition the desired results would be achieved and the Hydrogen injection rate would be lower. The results of experiments in different concentrations of dissolved Oxygen are shown in Figure (5). For example when the input of dissolved Oxygen is 0.5 ppm, the time needed to reach to the desired point of 5 ppb of remaining DO is short even at the contact time of 30 seconds.

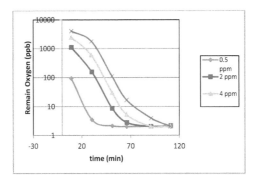

Figure 5. The effect of different concentrations of dissolved Oxygen

4. Conclusion

The operation condition is one of the most important parameters that influence on the efficiency of the catalytic reduction of dissolved Oxygen. The advantage of this process is its operation at moderate temperature and low pressure. The role of pressure is helping for Hydrogen solution in the system. Also the Hydrogen pressure prevents the formation of two phase flow after the mixing column. It is possible to reduce dissolved Oxygen by means of increasing the water flow rate at low inlet Oxygen level. In this condition the process would be more efficient.

Also, The contact time and Oxygen concentration are the most important parameters that influence on the efficiency of the catalytic reduction of dissolved Oxygen. In steam power plants that dissolved oxygen levels should be less than 5ppb, time is at least about 45 seconds. Also In a fixed volume of catalyst, at low amount of dissolved Oxygen in water inlet, it would be possible to increase water flow rate and decrease the Hydrogen injection and therefore reach to desired outlet Oxygen concentration.

Author details

Adnan Moradian[1], Farid Delijani[2] and Fateme Ekhtiary Koshky[2]

1 Ministry of Power, Niroo Research Institute,Tehran, Iran

2 Ministry of Power, East Azarbayjan Power Generation Management Co., Thermal Power Plant of Tabriz, Tabriz, Iran

References

[1] Martinola, F. B, & Thomas, P. Saving Energy by Catalytic Reduction of Oxygen in Feedwater", 45th International Water Conference, Eng. Soc. Pittsburgh,PA, ((1980).

[2] Chua, I, Tai, M. S. L, Li, K, Zhang, H, Sourirajan, S, Ng, W. J, & Teo, W. K. Ultrapure water quality and production for industrial uses", Journal of The Institution of Engineers ((1993).

[3] Li, K, Chua, I, Ng, W. J, & Teo, W. K. Removal of dissolved oxygen in ultrapure water production using a membrane reactor", Chem. Eng.Sci. (1995). , 50(1995), 3547-3556.

[4] Tan, X, Capar, G, & Li, K. Analysis of dissolved oxygen removal in hollow fibre membrane modules: effect of water vapour", Journal of Membrane Science (2005). , 251(2005), 111-119.

[5] Li, K, & Tan, X. Development of membrane-UV reactor for dissolved oxygen removal from water", Chem. Eng. Sci. (2001). , 56(2001), 5073-5083.

[6] Moon, J. S, Park, K. K, Kim, J. H, & Seo, G. Reductive removal of dissolved oxygen in water by hydrazine over cobalt oxide catalyst supported on activated carbon fiber", Appl. Catal. A-Gen. (2000). , 201(2000), 81-89.

[7] Butler, I. B, Schoonen, M. A. A, & Rickard, D. T. Removal of dissolved oxygen from water: A comparison of four common techniques",Talanta (1994). , 41(1994), 211-215.

[8] Ito, A, Yamagiwa, K, Tamura, M, & Furusawa, M. Removal of dissolved oxygen using non-porous hollow-fiber membranes", J. Membrane Sci. (1998). , 145(1998), 111-117.

[9] Tan, X, & Li, K. Investigation of novel membrane reactors for removal of dissolved oxygen from water",Chem. Eng. Sci. (2000). , 55(2000), 1213-1224.

[10] Suppiah, S, Kutchcoskie, K. J, Balakrishnan, P. V, & Chuang, K. T. Can." Dissolved oxygen removal by combination with hydrogen using wetproofed catalysts", J. of Chem. Eng., 66, ((1988).

[11] Moradian, A, & Delijani, F. F.Ekhtiary Koshky, " The Effect of Contact Time and con-
     centration of Oxygen on the Efficiency of the Catalytic Reduction of Dissolved Oxy-
     gen", J. of Middle-East journal of scientific research., (2012). , 11(2012), 1036-1040.

[12] LANXESS Energizing Chemistry: wwwlanxess.com

Chapter 6

Environmental Aspects of Coal Combustion Residues from Thermal Power Plants

Gurdeep Singh

Additional information is available at the end of the chapter

1. Introduction

Electricity is an essential need of any industrial society and no nation can progress without adequate supply of power. Growth in its demand during the past decades has been phenomenal and has outstripped all projections. In India also, there has been impressive increase in the power generation from a low capacity of 1330 MW in 1947 at Independence to about 81,000 MW at end of March 1995[1] and at about 2,05,340 MW at the end of June 2012 [2]. However, despite this substantial growth there still remains a wide gap between demand and supply of power which is expected to worsen in the years and decades to come. We already are experiencing shortage of nearly 8% of the average demand and 16.5% of peak demand. Moreover, with the quantum jump expected in demand for power in the future due to rapid industrialization and changing life styles of populace as a result of economic liberation, shortages shall further increase unless immediate steps are taken to increase power production.

1.1. The energy scene

Per capita consumption of energy in India is one of the lowest in the world. India consumed 570 kg of oil equivalent (kgoe) per person of primary energy in 2009-10 compared to 1090 kgoe in China and the world average of 1,688 kgoe [3]. India has declared to be moving towards energy security and independence in a few decades with ambitious energy generation targets. Total installed capacity source wise for June 2012, is give below in Table 1 and Figure 1 [2].

Source	Total Capacity (MW)	Percentage
Coal	116,333.38	56.65
Hydroelectricity	39,291.40	19.13
Renewable Energy Source	24,832.68	12.09
Gas	18,903.05	9.20
Nuclear	4780	2.32
Oil	1,199.75	0.58
Total	2,05,340.26	100

(Source: Central Electricity Authority, New Delhi, June, 2012)

Table 1. Total Installed Capacity of Electricity Generation from Various Sources

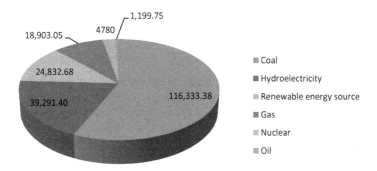

Figure 1. Total Installed Capacity of Electricity Generation from Various Sources (*Source: Central Electricity Authority, New Delhi, June, 2012*)

The total installed electric power generation capacity, public and private sector combined, is 2,05,340.26 MW only. Of the total installed capacity, thermal power is highest at 116.333.38MW representing almost 57%, the remaining being either gas based, oil based, hydro or nuclear [4]. Coal has been identified the mainstay fuel for thermal power generation in India and will remain so for another two decades at least.

1.2. Coal reserves and demand scenario

The Integrated Energy Policy Committee set up by GOI in 2004 emphasized the continuing dominance of coal in Indian Energy Security over the next 25 years [5]. India has vast reserves of coal and it is expected that coal remains a prime source of energy through the early part of 21st century The coal reserves in India up to the depth of 1200 meters have been estimated by the Geological Survey of India at 285.86 billion tonnes as on 1.4.2011 [6].. Only 12.5% of the resource is coking type and remaining 87.5% is non-coking or thermal coal. Coal deposits are mainly located in Jharkhand, West Bengal, Orissa, Chattrisgarh, Maharastra, Madhya Pradesh and Andhra Pradesh [6]. About 75% of the coal produced in the country is used for thermal power generation [7]. The coal demand is expected to rise to 2600 million tones at present level of consumption trend. The country's estimated demand of coal assessed by Planning Commission is 772.84 million tonnes for 2012-13 [6]. Majority of this demand will be met by our own resource of which more than 85% are high ash non-coking variety. Rapid increase in India's future power generation to meet the growing demand for domestic and industrial uses will be based on coal.

1.3. Coal combustion residues and their types

The process of coal combustion results in the generation of coal combustion residues (CCRs). India has diverse quality of coal reserves, which contain 30-55% ash. Annual production of CCRs has already crossed 120 MT and with the present rate of growth (8-10%) of power generation it is expected that this figure would be 175 MT by 2012 [8].

The quantum of CCRs generated in India is a matter of utmost concern. Out of the total CCRs generated, only 41% is being utilized [9]. Utilization that was around 1% in 1994 has shown a sharp increase in ten years period. However, the 41% utilization is still low as compared to more than 50% utilization in some other countries. One of the main reasons for the improper utilization of CCRs in India is its poor availability in usable forms and lack of characterization.

Coal combustion residues (CCRs) include materials that are left over after the burning of coal. They are fly ash, bottom ash, boiler slag and flue gas desulfurization (FGD) materials (wet or dry). The process of ash formation is influenced by the chemical composition of mineral matter in coal and its thermal properties. Other important factors include interaction between the inorganic and organic constituents of the particles and the physical and chemical environment through which these particles pass in the furnace. The distribution between the bottom ash and fly ash fraction is a function of the coal type, the boiler type and the type of boiler bottom.

1.3.1. Fly ash

Fly ash is one of the incombustible mineral residues produced from the combustion of pulverized coal in the boilers. The particle size of the fly ash is varied between less than 1 to 100 μm. The fly ash particles are removed with the help of flue gases from the stack and generally captured by electrostatic precipitators and any other air pollution control equipments such as bag houses or wet scrubbers [10].

Classification of Fly Ash: According to ASTM C 618 (American Society for Testing and Materials), fly ash is classified into two categories such as Class C and Class F fly ash. Both the type of ash contains SiO_2, Al_2O_3 and Fe_2O_3. Class C fly ash contains greater than 50% and Class F fly ash contains approximately 70% of these oxides. The average composition of Class C and Class F fly ash *is shown in Table 2.5* [11].

Class C Fly Ash: Class C fly ash is produced from the burning of sub bituminous coal or lignite. It has pozzolanic properties and self-cementing properties. Class C fly ash will harden and gain strength over time in the presence of water. Alkali and sulfate (SO_4) contents are generally higher in Class C fly ashes. It has higher lime (CaO) content (>10%) than Class F fly ash (<10%).

Class F Fly Ash: Class F fly ash is mainly produced by the anthracite and bituminous coal. It has pozzolanic properties and less than 10% lime content. Class F fly ash requires a cementing agent, such as hydrated lime or quicklime with the presence of water in order to produce cementitious materials.

Oxide	Class C (Wt%/std)	Class F (Wt%/std)
SiO_2	17.6 ± 2.7	52.5 ± 9.6
Al_2O_3	6.2 ± 1.1	22.8 ± 5.4
Fe_2O_3	25.2 ± 2.8	7.5 ± 4.3
CaO	"/>10	<10
MgO	1.7 ± 1.2	1.3 ± 0.7
Na_2O	0.6 ± 0.6	1.0 ± 1.0
K_2O	2.9 ± 1.8	1.3 ± 0.8
SO_3	2.9 ± 1.8	0.6 ± 0.5
LOI	0.06 ± 0.06	0.11 ± 0.14
Moisture	0.33 ± 0.35	2.6 ± 2.4

(Source: Scheetz et al., 1997)

Table 2. Composition of Class C and Class F Fly Ash

1.3.2. Bottom ash

Bottom ash is generally gray to blackish in colour. It is angular shaped and has a porous surface structure. Approximately 15-20% of total ash is bottom ash. It does not have any cementitious properties due to its larger size. Bottom ash consists of aggregated ash particles formed in pulverized coal boilers that are too large to be carried in the flue gases and fall through open grates to an ash hopper at the bottom of the boiler or impinge on the boiler walls [12]. Bottom ash is removed from the bottom of the boiler either in dry or wet state and is transported by the pipeline to handling system.

1.3.3. Pond ash

After the generation of fly ash it is transported to the specific area termed as ash pond. Generally pond ash contains fly ash and bottom ash mixture in the ratio of 80:20. About 20-40 cubic meters water is require for the transportation of 1 tonne of ash. The coarser particles are generally settled down due to the action of gravity. Pond ash creates respiratory diseases and reduction of visibility during the time of summer. That's why water is spraying over the ash pond surface to control the air pollution.

1.3.4. Boiler slag

Boiler slag is made up of black, hard and angular particles that have a smooth and glass like appearance. It is the molten ash collected at the bottom of the cyclone and slag tap that is mixed with water. When the molten slag comes in contact with the water, it fractures, crystallizes, and forms pellets. Boiler slag is mainly used for various applications such as fill material for structural applications, component of blasting grit and raw material in concrete products etc.

1.3.5. FBC (Fluidized Bed Combustion) ash

The FBC is a technology employed to reduce the amount of sulphur released into the atmosphere while burning sulphur-rich coal. FBC ash is formed when coal is burnt in the presence of crushed limestone as a bedding material in fluidized form. It consists of unburned coal, ash, and spent bed material. SO_2 is converted to $CaSO_4$ through its reaction with the limestone during the time of combustion. Approximately lower temperature (815 - 870⁰C) is require for fluidized bed furnace than the conventional coal fired furnaces (1400 -1600⁰C). FBC technology has become extensively used for reducing SO_2 emissions from power plants due to its low cost [13].

1.3.6. FGD (Flue Gas Desulfurization) ash

FGD ash is the solid residue that results from a variety of processes used to control SO_2 emissions from boiler stacks. Apart from over 95% SO_2 removal capacity, this technology can also reduce the emission of other gases like hydrogen chloride and sulphur trioxide [14]. FGD ash generally produced in the scrubber by the reaction of limestone with SO_2 to form calcium sulfite. This calcium sulfite further oxidizes to form calcium sulfate (95% pure).Both of the compounds are produced in the scrubbers in wet form and then dried and processed for handling.

1.4. Implications of coal ash generation

Generally high ash coals are available for thermal power generation. Solid waste from combustion, mainly fly ash removed with electrostatic precipitators or bag houses and bottom ash collected in the boiler, have become an important environmental problem because of their high volume and physical and chemical characteristics. As per the available estimates the production of coal ash in India including both fly ash and bottom ash is about 120 million tones per annum which is likely to touch 200 million tons per annum by 2015 A.D [15]. Due to high ash content in Indian coal, emphasis is being laid now on setting up coal fired Super, Ultra and Mega Power Projects on pithead itself in order to minimize cost associated with bulk transport of ash laden coal [16]. The installed capacity is expected to increase to about 300,000 MW by 2017. Majority of these additions would be coal based which will further add to the burgeoning problems of ash disposal. A thermal power plant generating 1000 MW of electricity produces about 1.6 million tones of coal ash annually of which 80% is fly ash. The management of large volume of coal ash produced in power plants is a real challenge for the nation [17].

The power plants must either find purposeful utilization of these ashes or alternatively dispose them off site. Although coal ash possesses beneficial properties, both physical and chemical. Still serious concerns related to health, safety and environmental risks involving air and water quality prevail in the mind of mine planners, operators, regulators and environment groups [18]. Thus the utilization of coal ash has drawn considerable concern and attention of scientists, technologists, environmental groups, government, regulators etc. Till the early 1990's only a very small percentage (3%) of the fly ash was used productively in India and the balance material was being dumped in slurry form in vast ash ponds close to power plants. The numbers of governmental and institutional actions taken since then have increased the ash utilization to 50% during 2010-2011 [19].

2. Environmental regulations for the fly ash utilization

To address the problem of pollution, caused by fly ash and to reduce the requirement of land for disposal of fly ash in slurry form in ash ponds, Ministry of Environment & Forests (MoEF) Government of India (GOI), has issued following notifications stipulating targets for utilization of the fly ash to achieve 100% utilization in phased manner [20].

2.1. Ministry of environment and forests — Notification

Published in the Gazette of India, Extraordinary, Part II, Section 3, Subsection (ii), New Delhi, the 6th November, 2008.

S.O. 2623 (E)._Whereas by notification of the Government of India in the Ministry of Environment and Forests number S.O. 763(E), dated the 14th September, 1999 (hereinafter referred to as the said notification) issued under sub-section (1) and clause (v) of sub-section (2) of section 3 and section 5 of the Environment (Protection) Act, 1986 (29 of 1986), the Central Government, issued directions for restricting the excavation of top soil for manufacture of bricks and promoting the utilisation of fly ash in the manufacture of building materials and in construction activity within a specified radius of one hundred kilometres from coal or lignite based thermal power plants [21]; (2) All coal and, or lignite based thermal power stations and, or expansion units in operation before the date of this notification are to achieve the target of fly ash utilization as per the given below:

S. No.	Percentage Utilization of Fly Ash	Target Date
1.	At least 50% of fly ash generation	One year from the date of issue of this notification
2.	At least 65% of fly ash generation	Two years from the date of issue of this notification.
3.	At least 85% of fly ash generation	Three years from the date of issue of this notification.
4.	100% fly ash generation	Four years from the date of issue of this notification.

(Source: Fly Ash utilization Amended Notification, MoEF, New Delhi, November 2009)

Table 3. Target of Fly Ash Utilization

The unutilized fly ash in relation to the target during a year, if any, shall be utilized within next two years in addition to the targets stipulated for those years. The balance unutilized fly ash accumulated during first four years (the difference between the generation and the utilization target) shall be utilized progressively over next five years in addition to 100% utilization of current generation of fly ash.

(3) New coal and, or lignite based thermal power stations and, or expansion units commissioned after this notification to achieve the target of fly ash utilization as per given below:

S. No.	Fly ash utilization level	Target Date
1.	At least 50% of fly ash generation	One year from the date of commissioning.
2.	At least 75% of fly ash generation	Two years from the date of commissioning
3.	100% of fly ash generation	Three years from the date of commissioning.

(Source: Fly Ash utilization Amended Notification, MoEF, New Delhi, November 2009)

Table 4. Target of Fly Ash Utilization

The unutilized fly ash in relation to the target during a year, if any, shall be utilized within next two years in addition to the targets stipulated for these years. The unutilized fly ash accumulated during first three years (the difference between the generation and utilization target) shall be utilized progressively over next five years in addition to 100% utilization of current generation of fly ash.

MoEF's Notification of 3rd November, 2009 i.e. [22]

i.   All thermal power stations in operation on the date of notification should have achieved the target of 60% fly ash utilization within two years from the date of notification i.e. by 3rd November, 2011; and

ii.   All new thermal power stations which have come into operation after the date of notification should have achieved the target of 50% of fly ash utilization within one year of their commissioning.

No person or agency shall within 50 kilometers (by road) from coal or lignite based thermal power plants, undertake or approve showing of mine without using at least 25% of fly ash on weight to weight basis, of the total stowing materials used and this shall be done under the guidance of the Director General of Mines Safety (DGMS) or Central Mine Planning and Design Institute Limited (CMPDIL).

Provided that such thermal power stations shall facilitate the availability of required quality and quantity of fly ash as may be decided by the expert committee referred in subparagraph (10) for this purpose.

8(ii) No person or agency shall within fifty kilometers (by road) from coal or lignite based thermal power plants, undertake or approve without using at least 20% of fly ash on volume to volume basis of the total materials used for external dump of overburden and same percentage in upper benches of back filling of opencast mines and this shall be done under the guidance of the Director General of Mines Safety (DGMS) or Central Mine Planning and Design Institute Limited (CMPDIL).

2.2. Strategies for coal ash utilization

Central Electricity Authority (CEA) has been working out the straggles on behalf of Government of India for effective utilization of the coal ash [15]. A large number of technologies have been developed for gainful utilization and safe management of fly ash under the concerted efforts of Fly Ash Mission of the GOI since 1994. As a result, the utilization of fly ash has increased to over 73 million tonne in 2010-11. Fly ash was moved from "hazardous industrial waste" to "waste material" category during the year 2000 and during November 2009, it became a saleable commodity. Fly ash utilization has started gaining acceptance, it being 55.79% during 2010-11.

The areas of concern include:

–Improving the collection efficiency of the ESP & of quality of fly ash generated.

–Need for development and implementation of systems for collection of classified fly ash, Certification of its quality for value addition and bulk environment friendly transportation options.

–Need for development of schemes for collection of dry bottom ash and its effective utilization.

–Need to develop energy efficient ash slurry pumps capable of handling dense ash slurry.

–Open trucks are used to transport fly ash for manufacture of building products. There is a need to develop efficient bulk transportation options for supply of fly ash from power plant to the end user. For 100% fly ash utilization at the generating stations

–Technologies are to be developed for demonstration of bulk utilization options of fly ash in roads and embankments, mine fills, sea erosion tetrapods, roller compacted concrete.

–Guideline standards to be developed to ensure quality assurance in value added products from fly ash, viz., bricks, blocks, pavers, kerbstones, tiles, etc.

–Development and application of high value added utilization of fly ash such as:

• Extraction of cenospheres (A cenosphere is a lightweight, inert, hollow sphere filled with inert air or gas, typically produced as a by-product of coal combustion at thermal power plants.)

• Extraction of titanium oxide, Alumina

• Development of composite materials, acid/fire resistant bricks /tiles

• Development of abrasion resistant materials

• Value added building materials

• Agriculture amendments, etc.

–**"Incubation centers"** should be set up for technology validation.

–**"Self sustaining technology demonstration centers"** to be established for technology propagation schemes.

–Encourage **"Industry–Institute interactions"** for entrepreneur development, awareness, training programmes and workshops.

–Induction of **"fly ash subject in academic curriculum"** of Engineering, Architecture, and Post Graduate Science Courses.

2.3. Modes of ash utilization during 2010–11

The major modes in which ash was utilized during the year 2010-11 along with utilization in each mode is presented in the pie diagram is given in Figure-2 below [15]:

MODE OF FLY ASH UTILIZATION DURING 2010-11

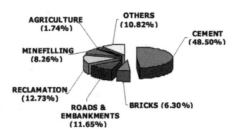

(Source: Central Electricity Authority, New Delhi, December, 2011)

Figure 2. Mode of Fly Ash Utilization during 2010-11.

It may be seen from above table that the maximum utilization of fly ash to the extent of 48.50% has been in Cement sector, followed by 12.73% in reclamation of low lying area, 11.65% in roads & embankments etc. The utilization of fly ash in mine filling was 8.26% and in making fly ash based building products like bricks, tiles etc was only 6.3%. These two areas have large potential of ash utilization which needs to be explored for increasing overall ash utilization in the country.

2.4. Progressive Fly Ash Generation and utilization during the period from 1996-97 to 2010-11

The progressive ash generation at coal/lignite based thermal power stations and its utilization for the period from 1996-97 to 2010-11 as per data received in CEA from power utilities is given in Figure 3 below [15]:

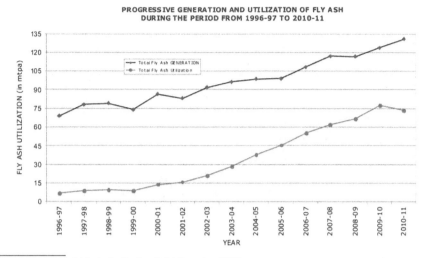

(Source: Central Electricity Authority, New Delhi, December, 2011)

Figure 3. Progressive Generation and Utilization of Fly Ash during the Period from 1996 – 97 to 2010 – 11.

3. Environmental aspects of trace elements leaching

There has been widespread concern of ground and surface water contamination due to trace elements leaching in and around the ash ponds. Leaching aspects of trace elements are presented here which are required to screen the coal combustion residues (Fly as-, bottom- & pond – ash} on environmental angle for its utilization.

3.1. Trace elements in fly ash

Studies of trace elements and the elements present in fly ash are distributed into the fractions of the fly ashes based on volatilization temperature [23]. It is found that elements appear to partition into three main classes.

Elements that are not volatilized and reported equally in both fly ash and bottom ash. These elements include Al, Ba, Ca, Ce, Co, Cu, Fe, Hf, K, La, Mg, Mn, Rb, Se, Si, Sm, Sr, Th, Ta, and Ti.

Elements that are volatilized on combustion and preferentially get adsorbed on the fly ash as flue gas cools down. These include As, Cd, Ga, Mo, Pb, Sb, Se and Zn.

Elements that remain almost entirely in the volatilised state tend to escape to the atmosphere as vapours. These are Hg, Cl and Br.

Fly ash is an alumino silicate glass consisting of the oxides of Si, Al, Fe and Ca with minor amounts of Mg, Na, K, Zn and S and various trace elements. The concentration associated with the ash may be either adsorbed on the surface of particle or incorporated into matrix [24]. A mechanism that appears to be common for all ashes during their formation is the condensation of metal and metalloid vapours on refractory core materials. As the ash particles and gas stream exist from the combustion chamber and proceed upto the flue gas, this results in locally higher concentrations of many trace elements at the surface of ash particles and accounts for the generally higher concentration of these elements as particle size decreases [25]. The association between trace elements and major elements/minerals may be an important factor in deter-mining the leachate composition of water in contact with ashes [26].

It is recognised that the health hazards and environmental impacts from coal fired thermal power stations result from the mobilization of toxic elements from ash [18, 24]. The large amount of ash that accumulates at thermal power plants, its possible reuse and the dispersion and mobilization of toxic elements from it, requires greater attention. Mobilization of various elements from the ash into the environment depends on climate, soils, indigenous vegetation and agricultural practices [27].

3.2. Leaching aspects of fly ash

The leaching characteristics of fly ash are mainly controlled by factors such as its chemical composition, mineralogy and morphology. Fly ash from thermal power plants vary in chemical composition not only from plant to plant but also within the same plant [26, 28]. Chemically fly ash consists of Si, Al, Mg, Ca, K, Ti and Fe in greater proportion with many trace elements as V, Mn, Cr, Cu, Ni, As, Pb, Cd and smaller quantity of various potential toxic elements. Chemical composition study shows mostly the presence of four major elements viz. aluminium, silicon, iron and calcium in the fly ash. Others such as potassium, magnesium, barium, cobalt, cadmium, zinc, molybdenum, lead etc. are present in traces [29]. Though in the traces, compared to the original coal, most of the elements are enriched in the fly ash, giving birth to the growing environmental concerns in the disposal and utilization environment due to release of trace/heavy elements metals [30].

Crystalline phases in fly ash as major constituent is quartz and others such as mullite, mag-netite, hematite etc. are also present as minor constituents. The mineralogy of fly ash refers to both amorphous and crystalline phases and other mineral fractions [29]. The leaching behavior of fly ash can be related to their mineralogical characteristics Most of the toxic elements reside on major phases and are easily available during leaching [26].

Morphologically, particles of fly ash may be irregularly shaped particles, solid spheres and cenospheres. Morphology of fly ash similarly, controls the leaching of toxic elements/heavy metals. Morphology of fly ash is related to the conditions of combustion and composition of inorganic materials in coal. Particle morphology is one of the most important morphological aspects of fly ash. In this particle size particle distribution also holds its importance. Finer particles mean more surface area and so more concentration of condensing elements. In fact, specific surface area of fly ash is determined by both particle size and particle morphology.

The leaching of fly ash is a time dependent phenomenon. The initial leaching of the fly ash can be characterized by the surface hydrolysis and the dissolution of reactive phases formed under high temperature combustion. A close examination of the leaching studies show a rapid early dissolution followed by a later, slower release of the elements. The water-soluble fraction of a combustion residue may reflect the early dissolution process in the natural environment. In fact, the early dissolution mainly involves the soluble salts or the oxides on the particle surface of the fly ash. So, the dominant features of the initial dissolution stage are a high dissolution rate and the solution chemistry being controlled by buffering components of the fly ash. At this stage of high dissolution rate, release of salts or heavy metals associated with surface phases occur. The long term leaching of the fly ash occurs primarily in the alluminosilicate glass and some acid-soluble magnetic spinel phases. These phases usually constitute the matrix of the fly ash. The dominant leaching features of the matrix phases are a low dissolution rate and phase alteration over a long period of time [26].

Present study on environmental characterization is in continuation with earlier studies, to evaluate leaching of trace elements from coal ashes from a few Thermal Power Stations situated in eastern India are presented in this study [26, 28, 31]. The possible water contamination is also envisaged through the leachate analysis from ash pond disposal sites in real life situation.

3.3. Leaching chemistry

Short-term leaching (shake) tests 24 and open column percolation leaching experiments were carried out on the fly ash samples to ascertain its leachate chemistry as briefly described below.

3.3.1. Strong acid digest test

This is short term leaching study and is carried out in the presence of strong acids to provide the available concentration levels of trace/heavy elements in the samples.

For the purpose of the experiment 0.5gm fly ash sample is taken in a conical flask and to it is added 10ml nitric acid and 2ml perchloric acid. Mixture is then heated till dryness on hot plate. The conical flask is covered during the process of heating by funnel. The dried residue in the conical flask is then boiled with 2ml HNO_3 and then filtered. This is repeated with distilled water, warmed up and filtered till no residue is left in the conical flask. The filter paper in the silica dish is kept in the muffle furnace and heated to 850°C. Silica dish is then allowed to cool with residue in it and residue is collected for further analysis. The filtered solution is stored in polypropylene beaker for further analysis (elemental).

3.3.2. ASTM shake test

This is the rigorous and short term leaching study. This test is run for twenty four hours and deionised water is used as the leaching medium. Shake tests can only be useful to a limited extent, and with variability in results in identifying the elements most likely to leach out of a material and to estimate the equilibrium constants for some of the reactions that takes place during the test. However, shake tests are poor indicators of the conditions that might be expected in the field [32].

In this test 80gm fly ash sample is taken and put in the measuring bottle and to it is added two litre of distilled water. The bottle with the sample is then taken in the rotary agitator and rotated for twenty-four hours and this way sample with water is agitated for thorough mixing to result leaching. The extraction is performed in triplicate on each fly ash sample and the three replicates are then mixed to get the composite leachate sample. The leachate so obtained is filtered using Whatmann No. 42 filter paper. The filtrate so collected is then stored in the polypropylene bottle for further analysis. Once the potentiometric analysis is over, few drops of 6N nitric acid are added to the leachate collected in polypropylene bottle to avoid contamination and the bottle stored to be used for further elemental analysis of the leachate by AAS.

3.3.3. 30–Day shake test

This test is similar to 24-hr shake test and is carried over a period of 30 days and is intended to indicate the solubility of elements that may reach equilibrium with the surrounding water more easily. This test is required by some regulatory agencies of some of the States in the USA.

3.3.4. Toxicity characteristics leachate procedure

The toxicity characteristic leachate procedure (TCLP) requires the use of an extraction fluid made of buffered acidic medium to run the test. For this the selection of the extraction fluid is made prior to conducting the test. Once the extraction fluid to be used in the test is determined, 40g fly ash sample is taken and then extraction fluid equal to twenty times the amount of sample taken is added in the zero head extractor under pressure. The system is tightly closed and then placed in an end-over-end rotary shaker for 18 hours, rotating at 30 ±2 rpm at a room temperature of about 25°C. Leachate after the said period of shaking is pressure filtered using 0.7 micron pore size filter paper.

3.3.5. Modified Synthetic Leachate Procedure (SLP)

This test is a modified SLP rather than a standard SLP. Here, unlike the standard SLP test which make use of a mineralized synthetic leaching medium prepared from deionized water, water from an actual field site (containing ions and impurities similar to those found in groundwater in an area of interest) is used. This test aims to detect the ion exchange reactions that can only be observed in the field like ion bearing water. This test is run for 24 hours.

3.3.6. ASTM column

In the ASTM column procedure, one pore volume of distilled water is forced through a packed column of fly ash each day in a saturated upflow mode. Leaching in this column is conducted under a nitrogen atmosphere and thus present leaching in an oxygen poor environment. In the field, this type of leaching would occur below the water table where there is low concentration of oxygen. One pore volume corresponds to the void space between the material grains in the test column. The rate of percolation of the leaching medium regardless of the hydraulic conductivity of the material is controlled by applying variable nitrogen pressure. The test is run for 16 days and leachate collected after 1, 2, 4, 8, 16 days of leaching [33].

3.3.7. Open percolation column experiments

In these experiments, deionized water is percolated through a packed column of fly ash in the presence of oxygen at a rate which depends on the natural permeability of the material. The open columns for leaching experiments are made of PVC pipe four inches in diameter and two feet in length. The column setup involved packing the coal ash material at optimum moisture and density conditions as determined by the Proctor test. The fly ash material is packed into the column in two inch lifts with a 2" x 2" wooden rod, about 4 feet long. Each packed layer is scarified, by lightly scraping the top of the packed layer with a long thin rod to ensure proper interlocking of the material. The top six inches of the column was left unpacked to allow for the addition and maintenance of the leaching medium. About 200 ml of leaching medium (de-ionized water) is added at the top of the column once every alternate day to maintain sufficient supply of water to the packed coal ash material. The top end of the column is exposed to the atmosphere and the bottom end is connected to quarter inch tubing. The columns discharged the leachates through this tubing into the 250 ml polypropylene beakers. The leachates are collected in these beakers and analyzed.

3.4. Elemental analysis of leachates

The leachate samples are filtered and acidified with 2 ml of nitric acid and then preserved in polypropylene sampling bottles. The samples are kept in a refrigerator until further analysis. Sodium and potassium were determined using flame photometer. Concentration levels of trace elements were evaluated using Atomic Absorption Spectrophotometer (AAS). Working/ standards solutions were prepared according to instructions given in the operation manual of the GBC-902 AAS [34]. Optimized operating conditions such as lamp current, wave length, slit width, sensitivity, flame type etc. as specified in the manual, are used for analysis of a particular element. -AAS standards are used for standardization and calibration of AAS. Three standards and a blank of the concerned element are used to cover the range 0.1-0.8 Abs. The calibration is performed by using the blank solution to zero the instrument. The standards are then analyzed with the lowest concentrations first and the blank is run between standards to ensure that baseline (zero point) has not changed. Samples are then an analyzed and their absorbance recorded. The calibration is performed in the concentration mode in which the concentration of sample is recorded. ICP-MS can be used to arrive at the precise and quick results wherever affordable.

4. Leachate analysis results

Comparative evaluation of short-term leaching (shake) tests is presented in Table 2. Acid digest data provides the available concentration levels of Trace elements in fly ash whereas in comparison to this shake tests resulted significantly lower concentration levels particularly in 24 hr. ASTM, 30 Day and SLP shake tests. TCLP leachate data however, gives rise to signifi-cantly higher levels as this involved leaching in slightly acidic buffered conditions and as such does not reflect the actual behavior. It may be emphasized that these shake test presented the accelerated leaching because of 1:20 solid to liquid (leachant) ratio. Nevertheless these short term (shake) leaching tests provide an immediate and rapid indication of leachable concentration levels of trace elements from fly ash.

Analysis of twenty two elements were carried out from each of the leachate samples collected from open column experiments and the observations are summarized in Tables 3 & 4 for fly ash, pond ash and actual ash pond leachates, respectively. It is noticed from the observations that the concentration of thirteen elements, namely, chromium, nickel, cobalt, cadmium, selenium, aluminum, silver, arsenic, boron, barium, vanadium, antimony and molybdenum were below the detection limit (.001 mg/l) in the entire study period. Among the other nine elements only calcium and magnesium were observed in the leachates throughout the study period while the concentration of other elements showed a decreasing trend to below detection limit (.001 mg/l). In the leachates from actual ash ponds, lead and manganese were found absent but iron, calcium, magnesium, sodium, potassium, copper and zinc were present throughout the study period.

A comparison of the concentration levels observed in the leachates of fly ash, pond ash and also leachates from actual ash pond disposal site with the permissible limits as per IS:2490, is presented in Tables 3 & 4 which indicates that the concentration levels of all the elements during the entire study period were either below detection limits (BDL) or below the permissible limits.

It can be inferred that no significant leaching occurs and toxicity is manageable with respect to trace elements both in the ash pond disposal site as well as in the open column leaching experiments. Further, analysis results of leachates from open column percolation experiments resemble closely with those of actual ash pond leachates. The physical set up of the open columns more closely resembles with because the flow of the leaching medium is influenced by gravity alone and the solid to liquid ratio is more close to the field situation. Hence, open column leaching experiments may be used in predicting the long term leaching behaviour that can be observed in the field. Fly ash leachates as generated from open percolation column leaching experiments and those from ash pond disposal site closely resemble and as such do not pose any significant environmental impacts in the disposal system. Overall, fly ash would not seem to pose any environmental problem during its utilization and/or disposal. Leaching pattern trace elements over three years open percolation column experiments is depicted in Figures 2-13 – which clearly reflect that trace elements leaching is not a significant concern and coal combustion residues can be appropriately utilized as these are established generally as environmentally benign material.

5. Concluding remarks

On the basis of the study of the leaching of trace elements from coal ashes, following conclusions can be drawn:

1. In the study period of about three (3) years there was practically no leaching of thirteen elements namely, chromium, nickel, cobalt, cadmium, selenium, aluminum, silver, arsenic, boron, barium, vanadium, antimony and molybdenum from all the ash samples.

2.   Out of the nine elements found in the leachates only calcium and magnesium were found to be leaching in the entire period. The leaching of other seven elements namely, iron, lead, copper, zinc, manganese, sodium and potassium was intermittent. The leaching of sodium and potassium practically stopped due to first flash phenomenon after 35 and 40 days, respectively. It is emphasised long-term leaching results should be considered to arrive at the environmental screening of such materials.

3.   The concentration of the elements in the leachates was invariably well below the permissible limits for discharge of effluents as per IS: 2490 and also for drinking water as per IS: 10500.

Parameter	Acid digest	TCLP	24-hr	30-D	SLP	SLP Blank
pH	--	4.29	6.22	6.26	7.08	7.06
Conductivity	--	3.56	0.096	0.099	0.075	0.085
TDS	--	1.78	48	63	51	58
Iron	82.41	0.089	0.045	0.05	0.029	0.038
Lead	BDL	BDL	BDL	BDL	BDL	BDL
Magnesium	7.579	7.512	3.15	4.50	2.4	2.8
Calcium	304.00	304.13	3.37	5.12	63	70
Copper	0.094	0.215	BDL	BDL	BDL	BDL
Zinc	0.276	2.140	0.020	0.025	0.180	0.185
Manganese	0.638	0.314	0.031	0.030	0.021	0.028
Sodium	54.60	1452	39.80	41.10	6	10
Potassium	7.60	8.10	2.80	5.36	3	4
Chromium	0.860	0.803	BDL	BDL	BDL	BDL
Nickel	0.118	0.112	BDL	BDL	BDL	BDL
Cobalt	BDL	BDL	BDL	BDL	BDL	BDL
Cadmium	BDL	BDL	BDL	BDL	BDL	BDL
Selenium	BDL	BDL	BDL	BDL	BDL	BDL
Aluminium	BDL	BDL	BDL	BDL	BDL	BDL
Silver	BDL	BDL	BDL	BDL	BDL	BDL
Arsenic	BDL	BDL	BDL	BDL	BDL	BDL
Boron	BDL	BDL	BDL	BDL	BDL	BDL
Barium	BDL	BDL	BDL	BDL	BDL	BDL
Vanadium	BDL	BDL	BDL	BDL	BDL	BDL
Antimony	BDL	BDL	BDL	BDL	BDL	BDL
Molybdenum	BDL	BDL	BDL	BDL	BDL	BDL
Mercury	BDL	BDL	BDL	BDL	BDL	BDL

BDL- Below Detectable Limit; Concentration of Elements in ppm; TDS in ppm; Conductivity in mmhos/cm

Table 5. Comparative Leachate Analysis Results of Shake Tests for Fly Ash

Parameter	Open Percolation Column Experiments Leachates			Ash Pond Leachate	(IS: 2490, 1981)
	Samples				
	FA#A	FA#B	PA		Inland Surface Water
pH	5.97-10.51	5.82-9.10	5.86-9.03	6.95-8.26	5.5-9.0
Conductivity	0.042-0.750	0.037-0.820	0.052-0.920	543-796	-
TDS	21-375	19-410	30-460	272-400	2100
Iron	BDL-0.740	BDL-1.220	BDL-1.369	0.89-1.983	-
Lead	BDL-0.420	BDL-0.396	BDL-0.490	0.121-0.462	0.1
Magnesium	BDL-15.53	0.039-38.00	0.065-44.00	17-29	-
Calcium	1.00-103.92	0.265-189.20	0.798-102-20	21-58	-
Copper	BDL-0.190	BDL-0.068	BDL-0.090	0.023-0.055	3
Zinc	BDL-0.380	BDL-0.372	BDL-1.529	0.295-1.763	5
Manganese	0.009-0.057	0.010-0.105	0.007-0.076	0.027-0.089	-
Sodium	3-56	3-49	3-47	19-43	-
Potassium	2-42	2-36	2-33	7-51	-
Chromium	BDL	BDL	BDL	BDL	2
Nickel	BDL	BDL	BDL	BDL	3
Cobalt	BDL	BDL	BDL	BDL	-
Cadmium	BDL	BDL	BDL	BDL	2
Selenium	BDL	BDL	BDL	BDL	0.05
Aluminium	BDL	BDL	BDL	BDL	-
Silver	BDL	BDL	BDL	BDL	-
Arsenic	BDL	BDL	BDL	BDL	0.2
Boron	BDL	BDL	BDL	BDL	2
Barium	BDL	BDL	BDL	BDL	-
Vanadium	BDL	BDL	BDL	BDL	-
Antimony	BDL	BDL	BDL	BDL	-
Molybdenum	BDL	BDL .	BDL	BDL	-
Mercury	BDL	BDL	BDL	BDL	0.01

BDL- Below Detectable Limit; Concentration of Elements in ppm; TDS in ppm; Conductivity in mmhos/cm

Table 6. Summary of the Leachate Analysis of Fly Ash from Thermal Power Stations# 1

Parameter	Open Percolation Column Experiments Leachates			Ash Pond Leachates	(IS: 2490, 1981)
	Samples				
	FA#1	FA#2	PA		Inland Surface Water
pH	4.98-9.92	4.38-8.90	4.71-8.92	7.2-8.58	5.5-9.0
Conductivity	0.060-0.962	0.070-0.848	0.036-0.973	645-892	-
TDS	30-481	35-424	30-487	320-445	2100
Iron	BDL-3.850	BDL-3.120	BDL-3.120	1.02-2.941	-
Lead	BDL-0.098	BDL-0.080	BDL-0.249	-	0.1
Magnesium	BDL-37.9	BDL-36.4	BDL-21.0	10-19	-
Calcium	1-87.6	2-72.2	2.12-48.0	18-46	-
Copper	BDL-0.094	BDL-0.088	BDL-0.052	0.011-0.047	3
Zinc	BDL-1.082	BDL-1.100	BDL-1.290	0.93-1.015	5
Manganese	BDL-0.099	BDL-0.092	BDL-0.069	-	-
Sodium	BDL-48	BDL-23	BDL-82	5-10	-
Potassium	BDL28	BDL-36	BDL-18	8-18	-
Chromium	BDL	BDL	BDL	BDL	2
Nickel	BDL	BDL	BDL·	BDL	3
Cobalt	BDL	BDL	BDL	BDL	-
Cadmium	BDL	BDL	BDL	BDL	2
Selenium	BDL	BDL	BDL	BDL	0.05
Aluminium	BDL	BDL	BDL	BDL	-
Silver	BDL	BDL	BDL	BDL	-
Arsenic	BDL	BDL	BDL	BDL	0.2
Boron	BDL	BDL	BDL	BDL	2
Barium	BDL	BDL	BDL	BDL	-
Vanadium	BDL	BDL	BDL	BDL	-
Antimony	BDL	BDL	BDL	BDL	-
Molybdenum	BDL	BDL	BDL	BDL	-
Mercury	BDL	BDL	BDL	BDL	0.01

BDL- Below Detectable Limit;

Concentration of Elements in ppm;

TDS in ppm;

Conductivity in mmhos/cm

Table 7. Summary of the Leachate Analysis of Fly Ash from Thermal Power Stations# 2

Overall, the fly ash samples from various Thermal Power Stations evaluated in this study were found to be environmentally benign and can be engineered for their bulk utilization particularly for mined out areas reclamation and for soil amendment for good vegetation.

The Centre of Mining Environment at ISM Dhanbad is currently engaged in evolving low technology high volume field demonstration to show that fly ash particularly fly ash can be disposed and utilized as fill material in an environmentally acceptable way in reclamation of abandoned mines [35, 36]. Fly ash has been successfully used as backfill in material in reclamation of mined out (goaf) area and at the top of the surface Helipad is set up, at Jamadoba Tata Steel Mining Area [37]. This has attracted a lots of public attention as a result of aesthetic and scenic value provided in the Eco-Park thus resulted. Similarly a considerable quantity of fly ash has also been utilised at Ghantotand OB Dumps and Damoda worked out opencast mine sites. At these sites trace elements leaching even after three years of monitoring does not seem to pose any environmental problem. With these encouraging results cooperative arrangements are being made by power utilities and mining authorities for utilisation of fly ash in reclamation of various mined out areas in SECL, MCL, NCL, BCCL, SCCL etc. [38].

Use of fly ash as backfill material for reclamation of mined out sites provide benefits such as easy availability, cheaper to transport because empty coal carriers returning from the power plant can "back haul" it to the mine site. From the standpoint of the power plant, this is essentially a waste material which requires large costs of handling and a disposal to comply with environmental regulations. From the environmental point, this waste material will go back to the same place where it was mined and use of this material serves as extra benefit to power plants. Studies are also in progress to use fly ash for agriculture development.

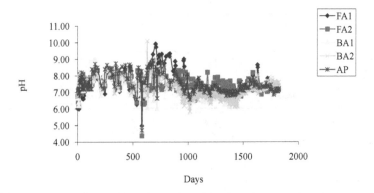

Figure 4. Open Column Leachate Analysis for pH

Figure 3.

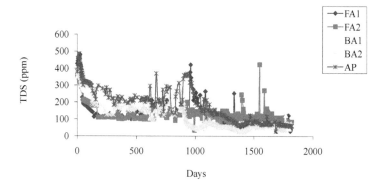

Days

Figure 5. Open Column Leachate Analysis for Conductivity

Figure 4. Open Column Leachate Analysis for TDS

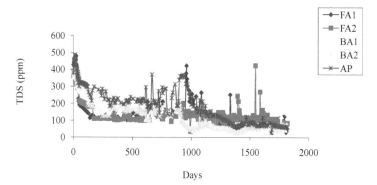

Figure 5.

Figure 6. Open Column Leachate Analysis for TDS

Figure 4. Open Column Leachate Analysis for TDS

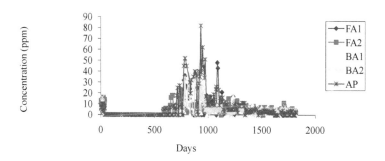

Figure 7. Open Column Leachate Analysis for Sodium

Figure 6. Open Column Leachate Analysis for Sodium

Figure 7.

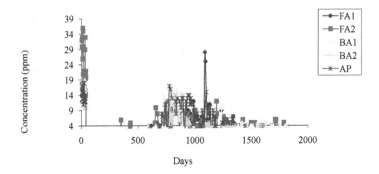

Figure 8. Opern Column Leachate Analysis for Potassium

Figure 8. Opern Column Leachate Analysis for Potassium

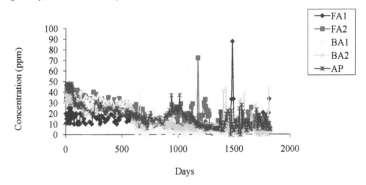

Figure 9. Open Column Leachate Analysis for Calcium

Figure 10. Open Column Leachate Analysis for Calcium

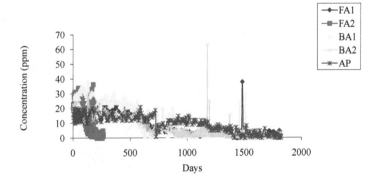

Figure 10. Open Column Leachate Analysis for Magnesium

Figure 12. Open Column Leachate Analysis for Magnesium

Figure 13.

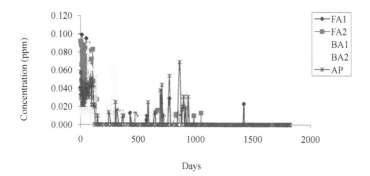

Figure 11. Open Column Leachate Analysis for Manganese
Figure 15.
Figure 14. Open Column Leachate Analysis for Manganese

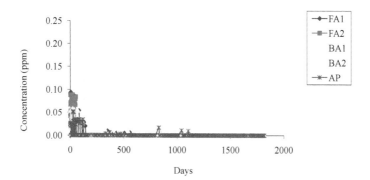

Figure 12. Open Column Leachate Analysis for Copper

Figure 16. Open Column Leachate Analysis for Copper

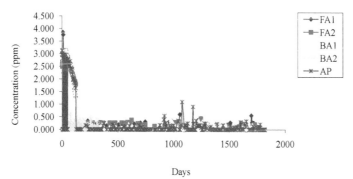

Figure 13. Open Column Leachate Analysis for Iron
Figure 18. Open Column Leachate Analysis for Iron

Figure 19.

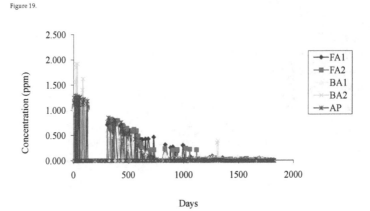

Figure 14. Open Column Leachate Analysis for Zinc
Figure 20. Open Column Leachate Analysis for Zinc

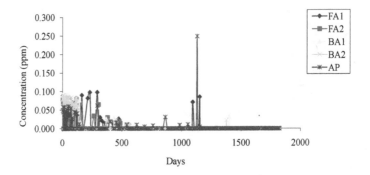

Figure 15. Open Column Leachate Analysis for Lead

Author details

Gurdeep Singh*

Department of Environmental Sciences and Engineering, Indian School of Mines, India

References

[1] India Energy ForumPower India Year Book- , 2005-06.

[2] Ministry of Power. http://www.cea.nic.in.

[3] Ministry of Petroleum and Natural GasAnnual Report, (2005).

[4] Trehan, A, Krishnamurthy, R, & Kumar, A. NTPC'S experience in ash utilization"
     Seminar on Fly Ash Utilization, New Delhi, March, (1996)., 26-27.

[5] Integrated Energy PolicyReport of The Expert Committee, Government of India
     Planning Commission, New Delhi, August (2006)., 1-137.

[6] Ministry of CoalAnnual Report, , 2011-12.

[7] Ministry of CoalAnnual Report, (2005).

[8] www.tifac.org.in

[9] Kumar, V, & Mathur, M. (2004). Fly ash- gaining acceptance as building, proceedings
     of the Seminar on recent trends in Building materials, Bhopal, India, , 55-65.

[10] Heidrich, C. (2003). Ash utilization-an Australian perspective. Proc. Of International
     Ash Utilization Symposium, Centre for Applied Energy Research, University of Ken-
     tucky.

[11] ScheetzBarry, E., Michael J. Menghini, Rodger J. Hornberger, Thomas D. Owens and
     Joseph Schueck, (1997). Utilization of fly ash. Proc. of Air and Waste Management
     Association. Toronto, Ontario, Canada.

[12] http://wwwadaa.asn.au

[13] Wang, H. L, Bolan, N. S, Hedley, M J, & Home, D. J. (2006). Potential uses of fluid-
     ized bed boiler ash (FBA) as a liming material, soil conditioner and sulfur fertilizer.
     In Coal Combustion Byproducts and Environmental Issues. Eds. Sajwan, K. S., Twar-
     dowska, I., Punshon, T. and Alva A. K. Springer, New York, USA. , 202-215.

[14] Srivastava, R. K. emissions-a review of technologies. Prepared for the U.S. Environ-
     mental Protection Agency, EPA/600/R-00/093.

[15] Central Electricity Authority (CEA)Report on Fly Ash Generation at Coal/Lignite
     Based Thermal Power Stations and its Utilization in the Country for the Year
     2010-11" New Delhi, December, (2011). , 1-18.

[16] Palit, A, Gopal, R, Dubey, S. K, & Mondal, P. K. Characterization and utilization of
     coal ash in the context of Super Thermal Power Stations. Proceedings International
     Conference on Environmental Impact of Coal Utilization, IIT, Bombay, (1991). ,
     154-155.

[17] Patitapaban SahuCharacterization of Coal Combustion By-Products (CCBS) for their
     Effective Management and Utilization, Thesis, (2010).

[18] Carloon, C. L, & Adriano, D. C. Environmental Impacts of Coal Combustion Resi-
     dues, Journal of Environmental quality 22 (1993). , 227-247.

[19] Kumari VibhaEnvironmental Assessment of Fly Ash in its Disposal Environment at F.C.I, Sindri, Jharkhand (India)", M. Tech Thesis, Submitted to Indian School of Mines Dhanbad, (1998).

[20] http://envfor.nic.in/

[21] Draft Notification on Fly ashS.O. 2623 (E), Published in the Gazette of India, Extraordinary, Part II, Section 3, Subsection (ii), New Delhi, dated 06 November, (2008).

[22] Fly Ash utilization Amended notification SO. 2804 (E), MOEF, New Delhi, dated 3rd November (2009).

[23] Bahor, B. F, & Gluskoter, H. J. (1973). Boron in illite as an indicator of paleosalinity of Illinois coals, JSP, , 43, 945-956.

[24] Natusch, D. F. S, Wallace, J. R, & Evans, C. A. Toxic trace elements: preferential concentration in Respirable particles". Science, (1974). , 183(4121), 203-204.

[25] Markowski, G. R, & Filbly, R. Trace element concentration as a function of particle size in fly ash from a pulverised coal utility boiler" Environmental Science & Technology, (1985). , 19, 796-800.

[26] Gurdeep Singh and Kumar SanjayEnvironmental evaluation on leaching of trace elements from coal ashes: a case study of Chandrapura Thermal Power Station". Journal of Environmental Studies and Policy (2000). , 2(2), 135-142.

[27] Gurdeep Singh and Sanjay Kumar GambhirEnvironmental evaluation of fly ash in its disposal" environment. Proceedings, International Symposium on Coal- Science Technology, Industry Business & Environment, Nov. 18-19, 1996. editors. Kotur S. Narsimhan & Samir Sen. Allied Publishers Ltd. New Delhi. (1996). , 547-556.

[28] Gurdeep Singh and Kumar SanjayEnvironmental evaluation of coal ash from Chandrapura Thermal Power Station of Damodar Valley Corp." Indian J. Environmental Protection. (1999). , 18(12), 884-888.

[29] Page, A. L, Elseewi, A. A, & Straughan, I. R. Physical and chemical properties of fly ash from coal fired power plants, Res. Review, (1979). , 71, 83-120.

[30] Gurdeep Singh and Kumari Vibha (1999). Environmental assessment of fly ash in its disposal environment at FCI, Ltd. Sindri, Poll. Res. , 18(3), 339-343.

[31] Jain, R. K. Environmental Assessment of Coal Combustion By-products of Burnpur Thermal Power Station, M. Tech. Thesis Submitted to Indian School of Mines, Dhanbad, (1998).

[32] Annual Book of ASTM Standards. (1990). 11.

[33] Annual Book of ASTM Standards. (1991). 04.

[34] Atomic Absorption Spectrophotometer Operating ManualAAS, GBC-902, Australia. (1990).

[35] Bradley, C. Paul, and Gurdeep Singh, Environmental evaluation of the feasibility of disposal and utilization of coal combustion residues in abandoned mine sites. Proceedings, First World Mining Environment Congress. Dec. 11-14, 1995. N. Delhi. Oxford & IBH Publ. Co. Pvt. Ltd. New Delhi. (1995). , 1015-1030.

[36] Bradley, C. Paul, Gurdeep Singh, Steven Esling, Chaturvedula and Pandal, H. The impact of scrubber sludge on ground water at an abandoned mine site. Environmental Monitoring and Assessment, (1998). , 50, 1-13.

[37] Singh, A. P. Environmental Evaluation of Coal Combustion Residues of Jamadoba Fluidised Bed Combustion Plant of TISCO with Special Emphasis on Stabilisation of Soil, M. Tech. thesis submitted to Indian School of Mines, Dhanbad, (2000).

[38] Bradley, C. Paul, Gurdeep Singh and Chaturvedula, S. 1995. "Use of FGD by products to control subsidence from underground mines: groundwater impacts." Proceedings International Ash Utilization Symposium Lexington, Ky, USA, October (1995). , 23-25

Chapter 7

Green Electricity from Rice Husk: A Model for Bangladesh

A.K.M. Sadrul Islam and Md. Ahiduzzaman

Additional information is available at the end of the chapter

1. Introduction

Bangladesh has reportedly over one hundred thousand Rice Mills - large, medium and small. They process paddy, using mostly the parboiling practice. The rice mills use the thermal energy of steam generated in boilers, which are fired by rice husk - a byproduct of paddy processing in the rice mills and which is globally well-known as very convenient source of dry biomass energy of reasonable heat value. In global context, annual rice husk production is 137 million tonnes whereas, in Bangladesh, about 9.0 million tonnes of rice husk is produced reported in 2011 [1]. In Bangladesh context, out of 37.08 million tonnes of total biomass produced from agro-residues, rice husk contributes about 26% by mass [2]. At present about 67-70% of rice husk is consumed for steam producing in rice mills [3, 4]. With a few exceptions, presently most of the 'boilers', used in these rice mills of Bangladesh are very inefficient. This results in a huge wastage of rice husk, which is an important source of biomass. Preliminary estimates are indicative that at least 50% of the rice husk produced could be saved and made surplus for its better use as input for small power generation [5]. Energy demand for rice processing is increasing every year due to the increased production of rice for ensuring food security of the population. More energy efficient rice parboiling boilers is to be suggested to replace the existing inefficient rice parboiling systems. So that the surplus amount of rice husks could be available for other sectors. If this husk is to be briquetted, then it will be an alternative fuel for replacing wood-fuel. Rice husk based co-generation system could be an alternative for large rice mills, or even for a cluster of small to medium rice mills. If the electricity were generated from rice husk then net carbon dioxide emissions could be reduced [6]. As a matter of basic dissemination strategy, applicable to most of the developing countries, initially the entrepreneurs/investors feel shy to invest in a relatively new area technology, as they lack the confidence on the technical functioning and financial/economic viability of ventures in such an area. Before launching a husk based power generation technology, it is important to know the production and distribution of available rice husk resource at different location. There are some different options to enhance the availability of rice husk such as increasing the un-parboiled process, improvement of thermal and combustion efficiency of existing husk fired boiler for parboiled process of paddy. Another important issue is the sustainable supply of rice husk in future for husk based power generation plant. Therefore, rice husk production and supply system is analyzed for electricity generation from husk in Bangladesh.

2. Methodology

There are main two sector of rice process viz. household level process and processed in rice mill. Husk obtained from household level process cannot be collected due to its wide spread localized. Therefore, husk from household level process is not considered as supply source of husk for electricity generation. Paddy processed in rice mill is considered the source of husk supply. There are several rice processing zones in the country. To conduct the investigation four major rice processing zones were selected to estimate potential husk available for electricity generation. The selected rice mill clusters are located at Dinajpur, Naogaon, Bogra, and Ishawrdi (Pabna). A survey questionnaire was used to collect the information on monthly quantity of paddy processed, amount of husk produced, amount of husk consumed for rice processing purposes and the surplus amount of rice husk (Survey questionnaire is shown in Appendix). Two hundred number of rice mills were surveyed in this study.

Another study is conducted to estimate the total electricity generation from rice husk in Bangladesh. LEAP (Long-range Energy Alternative Planning System) tool is used to make three different scenarios for consumption of energy. The basic parameter for estimating scenarios data are shown in Table 1.

Activity level	Reference scenario	Scenario-1
Paddy processed in mill	70%	90%
Parboiled processed paddy	90%	90%
Unparboiled processed paddy	10%	10%
Traditional parboiling	100%	5%
Improved parboiling	0	95%

Table 1. Assumption of activity levels at different scenario

3. Results and discussion

3.1. Available husk at rice mills

Annual paddy processing capacity and surplus amount of husk from the four selected rice processing zones are analyzed. The annual paddy processing capacities were estimated to be

936923, 394976, 1366980 and 925048 tonne per year for Pulhat (Dinajpur), Bogra, Naogaon and Ishawrdi cluster, respectively. Total paddy processed in these clusters was 3.62 million tonne. Amount of surplus husk for selected rice processing zones were estimated to be 82904, 586925, 192551 and 121209 tonne per year for Pulhat, Bogra, Naogaon and Ishawrdi cluster, respectively. Total surplus of husk were 455356 tonne per year in the selected four clusters.

3.2. Electricity generation capacity based on rice husk energy at the study areas

Electricity generation from rice husk depends on the availability of raw material and the technology for conversion rice husk to energy. For a steam turbine power plant consumption of rice husk is 1.3 kg per kWh electricity as reported by Singh [7]. On the other hand for a gasification power plant consumption of rice husk is 1.86 kg per kWh electricity generation [8]. Based on this assumption the capacity of electricity power generation was estimated for the study areas (Table 2). The potential power capacities is estimated to be 41450 kW in four selected zones considering the steam turbine technology, whereas the corresponding power capacities is estimated to be 29050 kW considering gasification technology.

Rice processing zone	Available husk from rice mill, '000' tonne/yr	Potential of electricity generation	
		Steam turbine plant kW	Gasification plant kW
		@1.3 kg husk/kWh$_e$ *	@1.86 kg husk/kWh$_e$ **
Dinajpur	82.90	7250	5150
Ishawrdi	121.21	10750	7500
Bogra	58.69	5200	3600
Naogaon	192.55	18250	12800
Total	**455.35**	**41450**	**29050**

*[8], **[7]

Table 2. Summary of electricity production from rice husk

Husk supply from rice mills is not uniform in quantity throughout the year. Sometimes the husk supply is surplus so that the husk is leftover after consumption and sometimes the husk supply is deficit than needed. Therefore, further analysis is done to see the surplus and deficit pattern of husk supply throughout the year. In this study it is estimated that the husk supply is surplus for 8 months and the quantity of surplus amount is 19428 tonne for Dinajpur. This surplus amount of husk is to be stored for the use to overcome the deficit of rice husk for rest of four months of the year. The deficit amount is 17956 tonne, therefore surplus amount of husk is sufficient to overcome the deficit amount for 7250 kW steam turbine power plant at Dinajpurt area (Fig. 1). Similarly 18540 tonne of surplus husk is sufficient to make up 18399 tonne shortage of husk for gasification power plant at Dinajpur (Fig. 2).

At Ishawrdi area, the husk supply is surplus for 9 months and the quantity of surplus amount is 16271 tonne. This surplus amount of husk is to be stored for the use to overcome the deficit of rice husk for rest of three months of the year. The deficit amount is 15806 tonne, therefore surplus amount of husk is sufficient to overcome the deficit amount for 10750 kW steam turbine power plant (Fig.3). Similarly 16432 tonne of surplus husk is sufficient to make up 15752 tonne shortage of husk for 7500 kW gasification power plant (Fig. 4).

At Bogra area, the husk supply is surplus for 9 months and the quantity of surplus amount is 2759 tonne. The deficit amount is 2474 tonne, therefore surplus amount of husk is sufficient to overcome the deficit amount for 5200 kW steam turbine power plant (Fig. 5). Similarly 3174 tonne of surplus husk is sufficient to make up 2335 tonne shortage of husk for 3600 kW gasification power plant (Fig. 6).

At Naogaon area, the husk supply is surplus for 9 months and the quantity of surplus amount is 23235 tonne. Whereas the deficit amount is 21299 tonne, therefore surplus amount of husk is sufficient to overcome the deficit amount for 18250 kW steam turbine power plant (Fig. 7). Similarly 22697 tonne of surplus husk is sufficient to make up 21477 tonne shortage of husk for 12800 kW gasification power plant (Fig. 8).

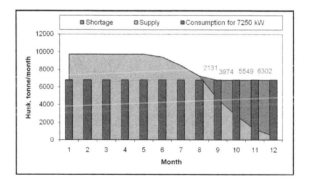

Figure 1. Potential production capacity of electricity at Pulhat for steam turbine plant

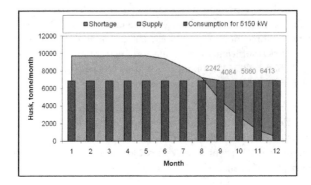

Figure 2. Potential production capacity of electricity at Pulhat for gasification plant

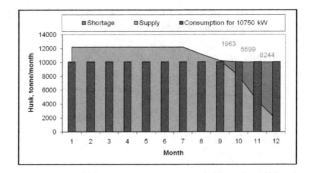

Figure 3. Potential production capacity of electricity at Ishawrdi for steam turbine plant

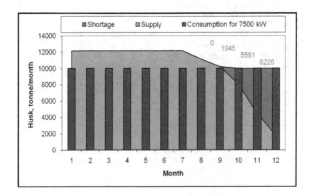

Figure 4. Potential production capacity of electricity at Ishawrdi for gasification plant

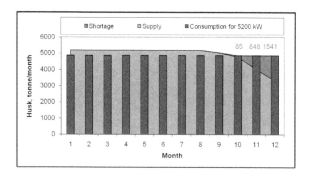

Figure 5. Potential production capacity of electricity at Bogra for steam turbine plant

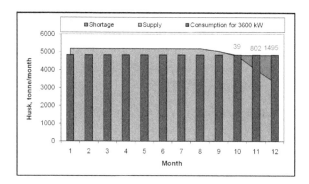

Figure 6. Potential production capacity of electricity at Bogra for gasification plant

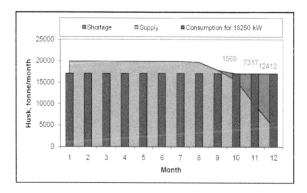

Figure 7. Potential production capacity of electricity at Naogaon for steam turbine plant

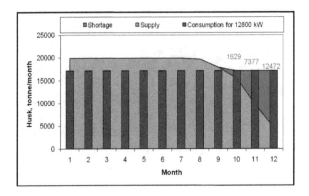

Figure 8. Potential production capacity of electricity at Naogaon for gasification plant

3.3. Projection of future rice production in Bangladesh

The production of rice is projected using the growth of 2.35%. The growth is calculated from the average growth of rice production data from 1971 to 2009. The projected rice production is increased to 71 million tonne in 2030 (Fig 9).

3.4. Rice processing in Bangladesh

All produced paddy is not processed for preparation of food grain. About 5% of total paddy is used as seed for using next growing season. Among the rest of paddy about 70% is processed in local rice mill. The rest 25% of paddy is processed in rural household level. Rice husk is produced as by product in rice mill at the rate of 20% of mass fraction of paddy. The rice husk is primarily consumed for steam producing in parboiling process and drying process of paddy. The rest amount of husk could be an important source of husk based electricity generation plant.

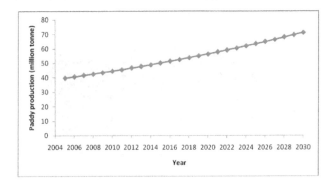

Figure 9. Raw rice production scenario in Bangladesh

3.5. Husk produced in rice mill

Amount of husk produced in rice mill depends on the amount of paddy processed in mill. Husk produced in rice mill could be increased by increasing the amount of processed paddy in mill. Share of paddy processed in mill is considered 90% of total paddy at the end of 2030 for scenario-1. As a result the husk production is increased from 9.96 million tonne (reference scenario) to 13.16 million tonne (Fig 10).

3.6. Husk consumed in rice mill

Husk is consumed primary for steam producing in rice parboiling process. More amount of parboiled paddy consumes more rice husk. At present about 90% of paddy is parboiled process. Therefore major amount of husk is consumed for parboiling of paddy. About 120 kg of husk is consumed for each tonne of paddy parboiled. The husk consumption for parboiling could be reduced from 120 kg to 49 kg for each tonne of paddy parboiled by replacing the traditional boiler with improved boiler. At present all boilers used are traditional type. In scenario-1, it is assumed that 95% of traditional boiler is to be replaced by improved boiler to reduce the husk consumption for parboiled process. The other way to reduce the rice husk consumption is reduction of parboiled coverage as well as enhancement of unparboiled process. The results of husk consumption for two different scenarios are shown in Fig 11. The husk consumption is reduction to 4.56 million tonne (Scenario-1) from 6.47 million tonne (reference).

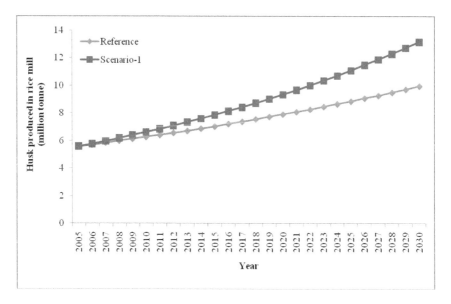

Figure 10. Rice husk production scenario in rice mill

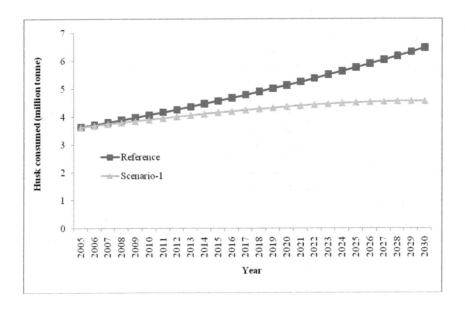

Figure 11. Rice husk consumption in rice mill at reference case ad scenario-1

3.7. Husk surplus after consumed in rice mill

The surplus amount of husk is calculated deducting the consumed amount of husk from total potential of husk (20% of paddy mass). The surplus amount of husk is increased to 8.60 million tonne (Scenario-1) from 3.49 million tonne (Reference) (Fig 12). The results show that there is huge potential to save rice husk by changing the technology for parboiling.

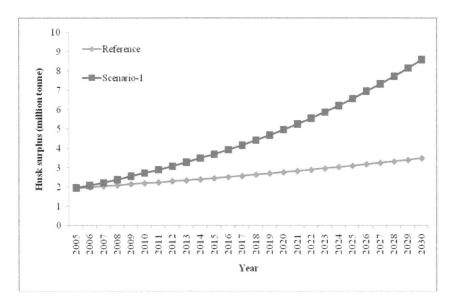

Figure 12. Rice husk surplus from reference case and scenarios-1

3.8. Potential of electricity generation from rice husk

Rice husk biomass easily can be used for electricity generation. Amount of electricity generated from rice husk depends on the amount of raw material and type of technology used for energy conversion. There are several technologies for electricity generation from biomass. In the present study steam turbine and gasification technology are considered for estimating the potential of electricity generated from surplus amount of husk from different scenarios. For a steam turbine power plant consumption of rice husk is 1.3 kg per kWh electricity as reported by Singh [7]. On the other hand for a gasification power plant consumption of rice husk is 1.86 kg per kWh electricity generation [8]. Based on this assumption the potential electricity generated from husk is presented in Fig 13 and Fig 14.

Potential electricity generated is estimated to be 1689 GWh (Reference), 2092 GWh (Scenario-1) in 2010 and these values will be increased to 2683 GWh (Reference) and 6612 GWh (Scenario-1) in 2030 for steam turbine technology use (Fig 13).

Potential electricity generated is estimated to be 1180 GWh (Reference) and 1462 GWh (Scenario-1) in 2010 and these values will be increased to 1875 GWh (Reference), 4621 GWh (Scenario-1) in 2030 for gasification technology use (Fig 14). Corresponding power capacities of husk based generation plant are shown in Fig 15 and Fig 16.

Electricity consumption for rice processing in Bangladesh is shown in Fig 17. Amount electricity consumed is estimated to be 812 GWh (Reference) and 891 GWh (Scenario-1) in 2010 and these values will be increased to 1292 GWh (Reference) and 2030 GWh (Scenario-1) in 2030.

The results show that amount of electricity generated from rice husk is sufficient to meet the electricity needed for rice processing purposes. Moreover, surplus amount of electricity could be served to national grid.

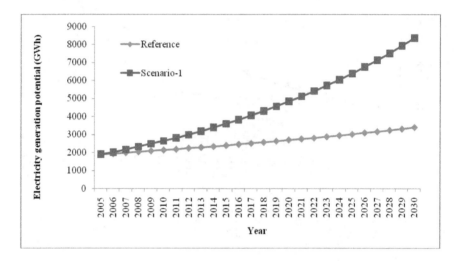

Figure 13. Potential of electricity generation from available rice husk at different scenario for steam turbine technology

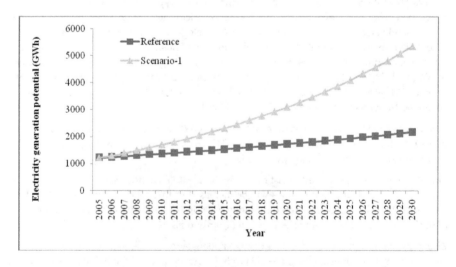

Figure 14. Potential of electricity generation from available husk at different scenarios for gasification based technology

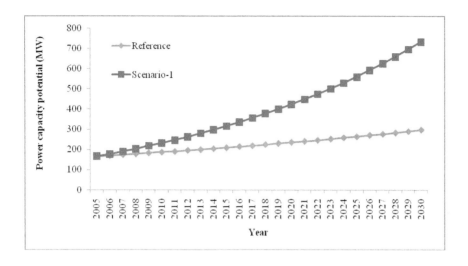

Figure 15. Potential of power capacity from available rice husk at different scenarios for gasification based technology

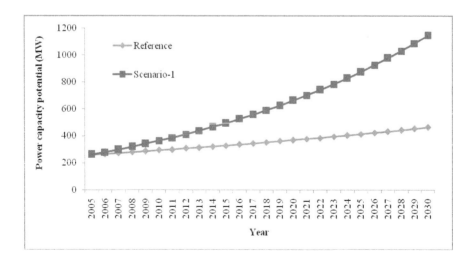

Figure 16. Potential of power capacity from available rice husk at different scenario for steam turbine technology

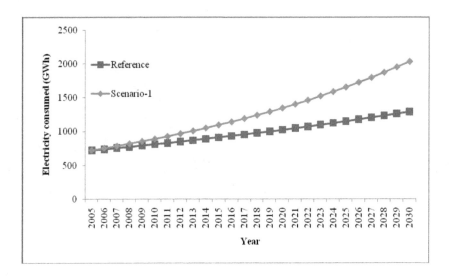

Figure 17. Electricity consumption for rice milling at different scenario

4. Conclusion

Power generation from rice husk is modern use of this waste material. The study results conclude that there is a potential to produce electricity from rice husk in selected rice processing zones of Bangladesh. It is estimated that 29 to 41 MW capacity generation plant can be installed in the selected areas based on the available surplus husk after their own consumption. In national context maximum 7682 GWh electricity can be generated from rice husk with a power capacity of 1066 MW in 2030. The amount of electricity generated from rice husk is sufficient to meet the electricity needed for rice processing purposes. Moreover, surplus amount of electricity could be served to national grid.

Apendix

QUESTIONNAIRE FOR THE SURVEY OF RICE MILL
WITH SPECIAL FOCUS ON
[PRODUCTION, OWN CONSUMPTION AND SURPLUS OF HUSK & AS WELL AS
DATA/INFORMATION ON ELECTRIC POWER CAPACITY]

A. GENERAL INFORMATION

01. Name of the Rice Mill.. Tel:......................................
02. Located in (mark√): ☐ Clusters / ☐ stand-alone [Location name:..]
03. District:..

B. TECHNICAL INFORMATION
B1. Paddy processing capacity

04. Types of clean rice produced (mark√): ☐ Parboiled   ☐ Un-parboiled   ☐ Puffed rice
05. Amount of paddy per batch: (a) Parboiled.............. bag   (b) Un-parboiled.............. bag
                             (c) Puffed rice.......... bag (d) Weight of paddy, kg/bag............
06. Average number of batch completed in a year:
         (a) Parboiled ....batch   (b) Un-parboiled..............batch   (c) Puffed rice..........batch
07. Total operating month/days in a year .... month / days

B2. Parboiling operation of paddy

08. Time required to complete the parboiling [wm× m¤úbœ Ki‡Z KZ mgq].......................... hours
09. Husk/bran mix husk consumption: (a) no. of bag/basket........... (a) weight of bag/basket ....kg
10. Motor capacity for blower feeding of husk: .... hp/kW

B3. Drying operation of paddy

11. Type of drying system (mark√): ☐ Sun drying   ☐ Mechanical   ☐ Hybrid (Sun + Mechanical)
12. Time required for sun drying: .... days
13. Husk required mechanical drying: (a) no. of bag/basket........... (a) weight of bag/basket ....kg

B4. Milling operation of paddy

14. Type of milling machine (mark√):   ☐ Modern rubber roll+polisher   ☐ steel sheller
15. (a) Motor capacity of steel sheller........... kW/hp   (b) No. of motor........................

B5. Amount of husk per batch of paddy processed

16. Amount of pure husk/bran mix husk   (a) No. of bag............... (b) weight........... kg/bag
17. Amount of bran   (a) No. of bag............... (b) weight........... kg/bag
18. Surplus of husk/bran mix husk   (a) No. of bag............... (b) weight........... kg/bag
19. Purpose of husk buyer:(a) Briquette making   (b) Feed (c) Poultry litter

C. Other observation (if any).

Name of Surveyor:..   Date:..................

Author details

A.K.M. Sadrul Islam[1*] and Md. Ahiduzzaman[2]

*Address all correspondence to: sadrul@iut-dhaka.edu

1 Department of Mechanical and Chemical Engineering, Islamic University of Technology (IUT), Gazipur, Bangladesh

2 Department of Agro-processing, Bangabandhu Sheikh Mujibur Rahman Agricultural University, Gazipur, Bangladesh

References

[1] FAOSTAT (2011) http://faostat.fao.org/site/339/default. aspx accessed on 20.8.11

[2] BBS (2009). Statistical yearbook of Bangladesh. Dhaka: Bangladesh Bureau of Statistics, Ministry of Planning.

[3] Ahiduzzaman, M (2007). "Rice Husk Energy Technologies in Bangladesh" Agricultural Engineering International: the CIGR Ejournal. Invited Overview No. 1. Vol. IX. January, 2007.

[4] Ahiduzzaman, M., Baqui, M. A., Tariq, A. S. and Dasgupta, N. (2009). Utilization of Rice Husk Energy for Rice Parboiling Process in Bangladesh. Intl. J. BioRes 6(2):47-79.

[5] Baqui, M. A., Ahiduzzaman, M., Khalequzzaman, M., Rahman, M. M., Ghani, J. and Islam, S.M.F. (2008): Development and Extension of Energy Efficient Rice parboiling Systems in Bangladesh. A Comprehensive Research Report submitted to German Technical Cooperation (GTZ).

[6] Ahiduzzaman, M. and Islam, AKM S (2009). Energy Utilization and Environmental Aspects of Rice Processing Industries in Bangladesh. Energies 2009, 2, 134-149; doi: 10.3390/en20100134.

[7] Singh, R. I. (2007). Combustion of Bio-Mass in an Atmospheric Fbc: An Experience & Study. Paper presented at the International Conference on Advances in Energy Research Indian Institute of Bombay, December 12-15, 2007

[8] Islam, K. (2008), Senior Adviser, SED project, GIZ, Dhaka.

Chapter 8

Feasibility of a Solar Thermal Power Plant in Pakistan

Ihsan Ullah, Mohammad G. Rasul, Ahmed Sohail,
Majedul Islam and Muhammad Ibrar

Additional information is available at the end of the chapter

1. Introduction

Pakistan has been facing an unprecedented energy crisis since the last few years. The problem becomes more severe throughout the year. The current energy shortage crisis has badly hit Pakistan's economy where hundreds of industries have closed due to lack of electricity to fulfil their requirements. The energy supply and demand gap has risen to 5000 MW [1] and is expected to rise considerably in the coming years as shown in Figure 1. Table 1 shows the existing installed power generation in Pakistan.

Pakistan has a huge potential in renewable energy especially solar energy to fill this gap if utilized properly. Pakistan, being in the Sun Belt, is ideally located to take advantage of solar energy technologies. This energy source is widely distributed and abundantly available in the country. Pakistan receives 4.45- 5.83 kWh/m^2/day of global horizontal insolation as an annual mean value, with 5.30 kWh/m^2/day over most areas of the country [2, 3]. This minimum level of solar radiation (4.45 kWh/m^2/day) is higher than the world average of 3.61 kWh/m^2/day [4] which shows that Pakistan lies in an excellent solar belt range. Pakistan has six main insolation measuring stations, namely Karachi, Islamabad, Lahore, Quetta, Peshawar and Multan and 37 observatories distributed fairly well over the entire country, recording sunshine hours as shown in Table A in Appendix. From the sunshine hours data it can be seen that most of the cities mentioned receive more than 250 sunshine hours a month.

2006 energy policy has resulted in few practical steps taken for utilizing the abundantly available solar resource in Pakistan. A Solar Water Heating System has been installed in a Leather Industry for first time in Pakistan. The system, using 400 m^2 Evacuated Collector tubes, provides heated water at 70 to 80 °C (at least 10 degree rise to the incoming water) to the already used boiler system, thus saving 33% of the cost. The Project was funded by Higher Education Commission (HEC) under University-Industry Technological Support Program (UITSP).

Solar water heating technology is relatively mature technology in Pakistan but its higher capital cost compared to conventional gas heaters has limited its use so far. This technology is widely recommended by a number of public sector organizations in northern mountains where natural gas is limited and difficult to supply. The solar water heaters are now being commercially produced in the private sector.

More than 2000 low cost solar cookers are used in Pakistan for cooking purposes. Similarly, solar dryers are used in Gilgit and Skardu (Northern part of Pakistan) to dry large quantities of fruits such as apricot and transport and sell them later in the urban areas, thus bringing economic prosperity to the area.

Fresh water unavailability in large parts of Baluchistan, Sind and southern Punjab is a critical issue. Two solar desalination plants consisting of 240 sills each with a capacity of 6000 gallons of seawater/day have been operational at Gawadar in Baluchistan province. A number of such schemes are under active consideration by local governments in Baluchistan and Thar [5].

The need for constructing solar power plants has been realized both at federal and provincial governments. The government of Sindh recently signed a Memorandum of Understanding with German company Azur Solar for building a 50 MW solar power plant at Dhabeji in District Thatta. The firm Azur Solar will initially set up a 60 kW solar power station to provide free electricity to backward 'goths' (Villages), schools and basic health centres of Badin.

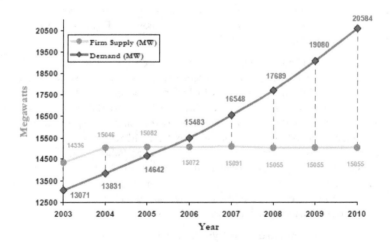

Figure 1. Generation capacity and demand forecast [6]

Both, technical and economical feasibility of a 50 MW solar thermal power plant using Stirling-Dish (SD) technology in Jacaobabad in Sindh Province of Pakistan is analysed in this chapter. The performance and environmental aspects of Stirling dish for power generation with and without solar energy is examined, discussed and compared. The solar data was collected from Bureau of Meteorology (BoM) of Pakistan and Life cycle cost analysis is performed to determine the economic feasibility of the solar thermal power plant. This study reveals that Jacobabad falls within the high solar energy belt and has minimum radiation energy of 4.45 kWh/m²/day (which is higher than the world average of 3.61 kWh/m²/day) to produce electricity, even during the low sunshine and cloudy days. The study concluded that solar thermal power plant in Pakistan is feasible using solar Stirling dish technology.

Type of Generation	Installed Capacity (MW)	Derated / Dependable Capacity (MW)	Availability (MW)	
			Summer	Winter
WAPDA Hydro	6444	6444	6250	2300
GENCOs	4829	3580	2780	3150
IPPs	6609	6028	5122	5402
Rental	285	264	250	250
Total	18167	16316	14402	11102

Table 1. Existing installed generation capacity in Pakistan [1]

2. Solar stirling engines

Solar thermal electric power generating systems have three different design alternatives, as follows;

- Power tower: a solar furnace using a tower to receive the focused sunlight

- Parabolic trough collector: focus systems that concentrate sun rays onto tubes located along the focal line of a parabolic shaped trough.

- Parabolic dishes: focus systems where sun light is reflected into a receiver at the dish's focus point [7, 8].

High optical efficiency and low start-up losses make dish/engine systems the most efficient (29.4% record solar to electricity conversion) of all solar technologies [9]. In addition, the modular design of dish/engine systems make them a good match for both remote power needs in the kilowatt range as well as hybrid end-of-the-line grid-connected utility applications in the megawatt range as shown in Figure 2 [9].

Solar Stirling engines can be classified into two categories;

- Free Piston Stirling engines: are those which have only two moving parts i.e. the power piston and the piston, which moves back and forth between springs. A linear alternator extracts power from the engine through power piston. Electricity is produced internally and therefore, there is no need for sliding seal at the high pressure region of the engine and no lubrication is required too [10].

- Kinematic Stirling engines: are those in which both the power piston and displacer (expansion and compression pistons) are kinematically (mechanically) connected to a rotating power output shaft.

Kinematic engines work with hydrogen as a working fluid and have higher efficiencies than free piston engines. Kinematic sterling engines have sealing problems and complicated power modulation. Sealing problems can be avoided by integrating a rotating alternator into the crankcase [10]. The power modulation can be sorted out by; (a) Varying the piston stroke (b) varying the pressure level of the working space [10].

Free piston engines have simple design as there is no connection between power piston and displacer. There is no need for working fluid make-up system being hermetically sealed as is required in the case of kinematic Stirling engines [11]. Free piston engines work with helium and do not produce friction during operation, which enables a reduction in required maintenance.

The solar Stirling engine is environment friendly as the heat energy comes from the sun and therefore almost zero emission. Similarly emissions from hydrocarbons combustion are very low as the fuel is burnt continuously at almost atmospheric pressure compared to the interrupted combustion in diesel and petrol engines. The quantities of CO produced and of unburnt hydrocarbons HCs are very low due to (i) the combustion of fuel in a Stirling engine occurs in a space surrounded by hot walls under adiabatic conditions and (ii) the latitude in the choice of air to fuel ratio. But unfortunately, the more efficient combustion of a Stirling engine results in relatively more CO_2 produced compared to an equivalent internal engine. Similarly the formation of NO_x are lower due to the short residence time of the gases at the high temperature, lower peak temperatures than in internal combustion engine and the continuous combustion. The emission of NO_x can be further reduced: (a) by recirculation of part of the flue gases along with incoming combustion air and (b) lowering the flame temperature. Stirling engine is intrinsically cleaner than all current heat engines in terms of emission of toxic or other polluting substances [11].

Figure 2. General description of Stirling EuroDish system [12]

3. Stirling-dish solar electric power generating system

Stirling-Dish (SD) systems are small power generation sets which generate electricity by using direct solar radiation. The capacity of a single unit is typically between 5 and 25 (50) kW$_{el}$. This size and the modularity of the single units qualify the Dish-Stirling system for very flexible applications. They are ideal for stand-alone or other decentralised applications [13]. The size of the solar collector for Stirling Dish is determined by the desired power output at maximum insolation levels (1 kW/m^2) and the collector and power conversion efficiencies. A 5 kW Stirling Dish system requires a dish of ca 5.5 m, in diameter, and a 25 kW system requires a dish of ca 10 m in diameter [10]. The schematic diagram of solar Stirling dish system is shown in Figure 3. The parabolic concentrator reflects the solar radiation onto a cavity receiver which is located at the concentrator's focal point. The heat exchanger (receiver) absorbs the solar radiation and thus heats the working gas (Helium or H$_2$) of the Stirling engine to temperatures of about 650° C. This heat is converted into mechanical energy by the Stirling engine. An electrical generator, directly connected to the crankshaft of the engine, converts the mechanical energy into electricity (AC). A sun tracking system rotates the solar concentrator continuously about two axes to follow the daily path of the sun to constantly keep the reflected radiation at the focal point during the day [13].

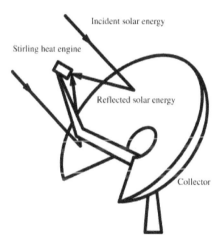

Figure 3. Schematic diagram of solar Stirling dish system [14]

4. Results and discussion

4.1. Technical feasibility

A SD solar electricity system of 25 kW with design characteristics shown in Table 2 was used for analysis in this study [10]. Stirling Dish has the ability to operate earlier and later every day and can also be functional in cloudy conditions when solar energy is < 2 kWh/m². Another advantage of SD is it can generate power between passing clouds due its low thermal inertia. Jacobabad's Latitude is 28°18′ N, Longitude is 68°18′ E and elevation is 55m. Jacobabad is among the hottest and the most arid areas of Pakistan, having weather conditions quite similar to those of a desert where direct normal insolation is high (1735 kWh/m²/year) [15] as shown in Table 5 coupled with low land cost, which is really cheap compared to international costs of land in Europe or USA.

Design Characteristics of a solar electricity system SD 25 kW	
Concentrator	
Glass area	91.01 m²
Aperture area	87.67
Focal length	7.45
Glass type	No. 82 commercial grade float. Thickness 0.7mm
Radius of curvature	599. 616, 667, 698"
Waviness	<0.6mr
Reflectivity	>90%
Module dimensions	11.89 mH, 11.28 W
Module weight	6.934 kg
Stirling engine (kinematic)	
Engine dry weight	225 kg
Displacement	380 cc
Engine dimensions	66 cm W, 71 cm H, 58 cm L
Number of pistons	4 double acting
working fluid	H_2 or He
working fluid pressure	20 MPa
Operating temperature	720°C
Power control	Fluid pressure
Cooling	Water/forced air fan
Output power	27 kW (max), 22 kW (rated)
Rated power efficiency	38-40%
Power conversion unit	
Weight	>680 kg
Alternator	Induction, 1800 rpm
Alternator efficiency	92-94 %
Electrical power	480 V, 60 Hz, three phase
Gross power rating	25 kW at 1000 W/m²
Peak net power efficiency	29-30%
Minimum insolation	250-300 W/m²
Dimensions	W=168 cm, H=122 cm, L=183 cm

Table 2. Design characteristics of 25 kW solar Stirling Dish [12]

NOAA Code	Statistics	Units	Jan	Feb	Mar	Apr	May	Jun	Jul	Aug	Sep	Oct	Nov	Dec	Average
0101	Temperature Mean Value	F	59.2	64.2	74.8	86.4	94.8	98.4	94.8	91.8	88.5	82	71.8	61.5	80.7
0109	High Temperature Mean Daily Value	F	72.7	77.4	88.3	100	110	112	105	101	98.6	95.5	86.2	75.4	93.5
0110	Low Temperature Mean Daily Value	F	45.9	50.9	61.3	72.1	80.1	84.9	84.6	82.9	78.5	68.5	57.4	47.7	76.9
0615	Precipitation Mean Monthly Value	Inches	0.1	0.3	0.4	0.1	0.1	0.2	1.5	1.1	0.5	0.1	0	0.2	0.4
0101	Temperature Mean Value	C	15.1	17.9	23.8	30.2	34.9	36.9	34.9	33.2	31.4	27.8	22.1	16.4	27.05
0109	High Temperature Mean Daily Value	C	22.6	25.2	31.3	38	43.1	44.3	40.6	38.2	37	35.3	30.1	24.1	34.15
0110	Low Temperature Mean Daily Value	C	7.7	7.7	16.4	22.3	26.7	29.4	29.2	28.3	25.9	20.3	14.1	8.7	19.95
0615	Precipitation Mean Monthly Value	mm	3.1	7.1	10.3	2	1.7	4.7	36.8	26.3	11.2	2.3	1.2	3.7	9.2

Table 3. Climate average weather data for Jacobabad [16]

The Jacobabad's 10 years sunshine hours data was provided by Pakistan Meteorological Department as shown in Table 4. It can be seen from Table 4 that there are very good sunshine hours throughout the year and the average sunshine hours is 9 for a month. It is slightly less than that of National Renewable Energy Lab NREL total sunshine hours [17] but even then it is highly suitable for a solar power plant.

JACOBABAD SUNSHINE HOURS PER MONTH											
Year	Jan	Feb	Mar	Apr	May	Jun	Jul	Aug	Sep	Oct	Nov	Dec
2000	213.3	248.1	240.8	273.7	364.1	361.2	330.0	307.4	297.8	308.5	241.6	245.2
2001	269.7	207.8	266.0	261.6	342.1	303.1	209.2	315.4	290.3	287.2	287.4	249.6
2002	238.5	207.4	242.0	263.2	330.7	342.4	300.6	240.6	281.7	300.1	243.3	239.9
2003	200.8	194.0	222.4	128.3	298.2	320.0	228.6	270.1	211.5	294.7	254.2	233.5
2004	200.8	194.0	255.0	128.3	262.2	275.6	270.3	256.3	237.5	318.2	294.5	238.1
2005	200.8	194.0	218.7	128.3	298.5	328.6	341.9	333.4	284.5	313.3	271.7	264.4
2006	216.0	193.9	251.8	247.4	341.9	331.4	312.7	251.2	307.7	310.9	221.4	203.2
2007	213.4	200.0	293.0	319.0	385.5	331.5	313.0	341.5	343.5	335.0	289.5	270.5
2008	171.0	254.5	279.0	273.5	359.0	288.0	323.0	341.5	328.0	315.5	289.0	258.5
2009	164.7	255.9	289.1	320.6	252.0	252.0	269.0.	372.0	281.2	309.5	285.3	247.4

Table 4. Jacobabad sunshine hour's data [3]

It is the hottest location in Pakistan as the temperature is around 45 to 50°C in summer and 10 to 25°C in winter as shown in Figure 4. Temperature data given in Table 3 also shows a low temperature mean value of 19.95°C, high temperature mean value of 34.15°C and Temperature mean value of 27.05°C very good hot conditions at Jacobabad [16], which are highly conducive for a solar thermal power plant.

Direct normal insolation data from National Renewable Energy Lab (NREL) USA, shown in Figure 5 and Table 5, was compared with the data from BoM (Pakistan) for assessment of the Jacobabad site. It can be seen that there is not a huge difference between the NREL data and the data received from Pakistan Bureau of Meteorology.

Month	Monthly DNI (kWh/m²)	Average Day Length (Hours)
January	114.42	10.4
February	127.37	11
March	149.65	11.8
April	165.61	12.7
May	169.83	13.4
June	177.66	13.8
July	168.32	13.6
August	131.59	13
September	157.78	12.2
October	154.77	11.3
November	115.63	10.6
December	102.98	10.2
Monthly Average	144.63	12
Annual Total	1735.56	

Table 5. Average Direct Normal Insolation data for Jacobabad [15]

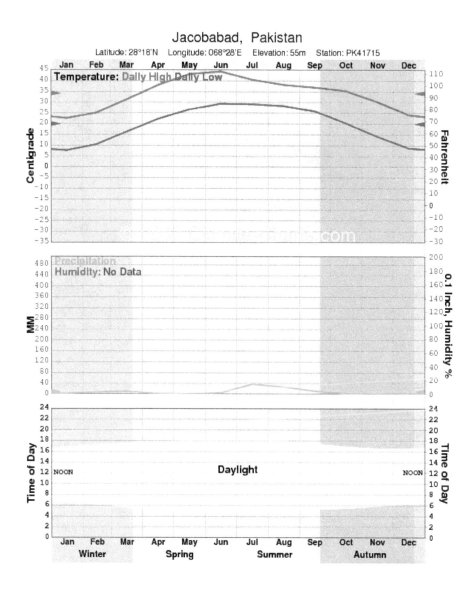

Figure 4. Jacobabad daily temperature – daylight chart [16]

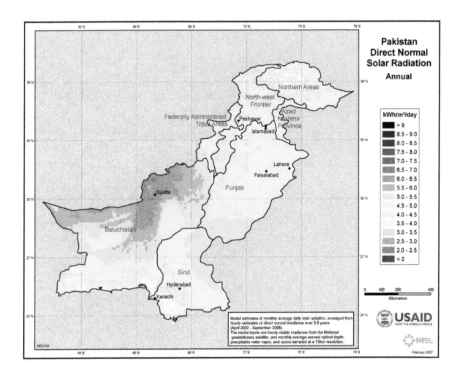

Figure 5. Pakistan annual direct normal solar radiation [17]

By comparing the sun shine hours of these cities available data, it was seen that sun shine hours in Jacobabad are more than other cities as shown in Table A1 of Appendix.

4.2. Economical feasibility

The selected economic indicators for this analysis are: Levelized cost of electricity, net present value (NPV) and total payback period (Tpb) [14].

$$C_L = \frac{I+O}{E_1 \sum_{t=1}^{n} \frac{1}{(1+k)}} [\$/kWh]$$

(1)

where I, O E$_1$, k and n are values of investment, operations and maintenance, energy produced by the plant in the first year (taken as the average annual energy produced), discount rate and number of years envisaged as the plant life time. Therefore, levelized cost is the total cash flows of a plant divided by the energy produced over the lifetime of a plant. It has been assumed that land is provided by the government. The operation and maintenance costs (O&M) cost in Pakistan for a solar thermal power plant is $ 0.001/kWh [18]. The energy inflation is considered to be negligible. We assume 60% sunshine hours are available throughout the year for electricity production to account for cloudy conditions and other miscellaneous factors. We shall use a discount rate of 10% for this analysis. The assumption and data used for the analysis is shown in Table 6, a [19]

Assumptions and data used	
Technical data	
Total power (MW)	50
Annual solar radiation (kWh/m^2)	1735
Annual generated electricity (MWh)	18970
Discount rate	10%
Lifetime (years)	25
System purchase price (US$/kWh)	4540[a]
Fixed cost	
Procurement of equipment (M$)	227
Transport & Installation (M$)	3[a]
Other costs (M$)	3[a]

Table 6. Assumptions and data

Net present value (NPV) which represents the discounted cash flow over the lifetime of a plant can in general be stated as

$$NPV = (1\text{-}T)\,(E-O) - I \tag{2}$$

Where T is marginal tax rate, E is income of the plant, O is O&M costs and I is initial investment.

Total payback period is the ratio of the initial capital investment I to the net income (difference between the value of the energy produced in the first year of operation, E_1, and the sum of the expenditure on operation and maintenance in the first year, O_1,

$$T_{pb} = I/(E_1 - O_1) \tag{3}$$

Quantity	Calculated Value
Value of investment	233 million US $
O&M Costs	0.475 million US$
Levelized electricity cost	24.4 cents/kWh
Sale price	60 cents/kWh
Net present value	22 million US$
Payback Period	23 years

Table 7. Summary of calculated quantities for a 50 MW solar thermal power plant

Solar dish technology can be a promising technology which can be deployed in Pakistan in small scale projects producing 25 kW electricity for remote areas, especially, in Sindh and Baluchistan and Federally Administered Tribal Areas (FATA), where there is enough solar potential for producing electricity through dish technology. Jacobabad has a great potential for a solar power plant based Stirling dish. The solar insolation is in the required limits which is more than 1700 kWh/m^2/year coupled with prolonged sun shine hours in those areas make this technology a viable option for producing electricity. Cost analysis shows that this technology is viable although it is expensive at the moment compared to fossil fuels based electricity. The prices will become competitive if the government of Pakistan withdraws the subsidies on fossil fuels and allocate it to solar based electricity. The calculations show that cost of electricity generation is higher than the existing water, natural gas and fossil fuels based generation which is 3 to 12 USD cent/kW [20] and therefore, government subsidy is necessary to make it competitive in the market. If government can provide subsidy on renewable power then it will attract more investment both at local and international level.

5. Conclusions

Pakistan has a huge potential for solar energy potential especially very high in Quetta and Karachi where a solar thermal power plant is viable. Government of Pakistan has to take strong policy and marketability measures that will establish and strengthen this environment friendly technology in the country. Solar power has very little impact on environment. This makes it one of the cleanest sources of power generation available to mankind. An operating solar power plant produces no air and noise pollution. Furthermore there is no hazardous waste produced in the production of electricity and it also does not require and transportable fuel. The use of solar electric systems is also known to reduce local air pollution. This results in the reduction in the use of kerosene and other fuels for lighting purposes. Solar power systems also help in the abatement of CO_2 gases.

Jacobabad is one the hottest place in the country and therefore, has the potential to have solar power plants installed there. Analysis for Stirling dish shows that it is feasible to install such systems there as the insolation is suitable for a solar power plant. The sunshine hours are also good coupled with good infrastructure i.e. transmission lines, natural gas and coal are in close proximity and easy access to national highway makes it an ideal place for a solar thermal power plant. Gawadar and Karachi ports are in the range of 500 to 700 km and also 132/220 kV transmission lines pass through the area for grid connectivity of the power produced by this plant. A positive net present value coupled with a reasonable payback period of 23 years indicates that this plant is a good option.

Appendix

LAHORE SUNSHINE HOURS DATA

	JAN	FEB	MAR	APR	MAY	JUN	JUL	AUG	SEP	OCT	NOV	DEC
2000	170.9	218.5	258.3	297.3	250.9	237.6	223.2	278.6	256.6	286.9	209.2	237.8
2001	185.8	234.1	277.5	247.4	308.0	244.9	217.2	284.4	263.2	253.3	250.5	198.3
2002	197.8	210.6	258.9	258.9	321.5	288.1	252.0	228.7	241.1	261.3	235.7	205.6
2003	111.9	204.8	234.9	279.7	319.8	245.4	254.0	230.5	236.9	277.0	345.6	225.0
2004	164.2	285.6	285.4	244.1	321.2	280.5	259.3	218.6	251.0	227.2	206.7	203.3
2005	210.3	150.7	220.7	271.9	285.4	292.1	206.9	258.5	238.5	285.6	255.1	227.3
2006	199.8	201.6	237.5	288.0	304.1	276.9	209.5	210.2	248.1	269.2	203.5	195.1
2007	234.9	138.1	260.7	308.5	298.8	250.2	236.4	226.2	223.3	284.7	215.8	217.3
2008	199.9	204.0	264.2	269.9	192.7	192.7	202.5	211.3	255.6	248.9	221.4	176.4
2009	185.2	220.3	238.5	276.2	310.5	308.7	252.9	228.0	273.5	217.5	147.1	189.1

ISLAMABAD SUNSHINE HOURS DATA

	JAN	FEB	MAR	APR	MAY	JUN	JUL	AUG	SEP	OCT	NOV	DEC
2000	141.1	197.9	255.9	293.5	316	268	235.6	273.6	259.5	258.9	185.7	183.6
2001	208.8	201.1	253.5	246.9	325.3	249.5	219.7	263.3	288.3	242.6	217.3	181.7
2002	202.2	155.4	252.1	259.2	323	307	317.7	181.9	267.1	245.7	201.7	191
2003	205.7	155	191	246.3	299.4	281.7	258.6	251.2	201.2	285.5	211.2	155.6
2004	115.3	224.9	252.6	169.9	313	275	270.5	224.2	248.6	221.2	202.5	157
2005	164.7	100.5	148.1	186.5	390.2	273.5	249.6	265.2	191.1	256.9	216.6	182.6
2006	142.7	154.7	209.7	246.9	315.1	264.1	192.5	208.8	242.8	245.7	163.9	146.1
2007	201.6	116.6	190.3	299.6	291.2	----	244.2	240.4	229.3	278	191.6	165.2
2008	175.5	205.6	245.6	307.8	224.1	224.1	228.7	231.4	244.3	244.4	255.9	199.7
2009	169	170.9	175.7	246.1	338.1	281.1	314.8	273.1	259.3	265.3	218.4	196.5

PESHAWAR SUNSHINE HOURS DATA

	JAN	FEB	MAR	APR	MAY	JUN	JUL	AUG	SEP	OCT	NOV	DEC
2000	148.3	225.1	204.0	275.3	277.2	250.2	253.2	270.4	253.3	258.4	194.5	131.5
2001	202.6	230.9	243.4	242.2	303.7	288.4	218.5	277.5	252.1	247.3	238.5	157.1
2002	204.1	155.8	222.7	246.7	305.0	280.3	285.8	189.8	243.4	248.1	203.3	140.2
2003	174.9	157.7	194.1	246.5	281.1	297.0	272.1	244.1	224.0	291.4	235.0	173.3
2004	135.6	227.8	259.1	214.2	314.5	303.0	249.3	272.6	249.4	227.3	213.3	137.8
2005	171.1	115.6	169.0	233.4	250.4	252.3	295.9	269.6	271.2	264.1	216.2	172.4
2006	148.6	156.1	172.2	256.5	289.5	----	124.7	148.6	254.2	222.2	163.8	124.5
2007	213.4	116.8	211.3	277.1	249.5	253.3	287.1	219.8	198.1	188.7	134.1	172.4
2008	150.2	173.8	197.6	255.1	248.6	248.6	226.2	221.2		221.6	177.8	176.3
2009	152.6	155.3	165.5	204.8	270.0	241.9	266.5	247.5	237.9	232.8	181.3	158.8

JACOBABAD SUNSHINE HOURS DATA

	JAN	FEB	MAR	APR	MAY	JUN	JUL	AUG	SEP	OCT	NOV	DEC
2000	213.3	248.1	240.8	273.7	364.1	361.2	330.0	307.4	297.8	308.5	241.6	245.2
2001	269.7	207.8	266.0	261.6	342.1	303.1	209.2	315.4	290.3	287.2	287.4	249.6
2002	238.5	207.4	242.0	263.2	330.7	342.4	300.6	240.6	281.7	300.1	243.3	239.9
2003	200.8	194.0	222.4	128.3	298.2	320.0	228.6	270.1	211.5	294.7	254.2	233.5
2004	200.8	194.0	255.0	128.3	262.2	275.6	270.3	256.3	237.5	318.2	294.5	238.1
2005	200.8	194.0	218.7	128.3	298.5	328.6	341.9	333.4	284.5	313.3	271.7	264.4
2006	216.0	193.9	251.8	247.4	341.9	331.4	312.7	251.2	307.7	310.9	221.4	203.2
2007	----	200.0	293.0	319.0	385.5	331.5	313.0	341.5	343.5	335.0	289.5	270.5
2008	171.0	254.5	279.0	273.5	359.0	288.0	323.0	341.5	328.0	315.5	289.0	258.5
2009	164.7	255.9	289.1	320.6	252.0	252.0	269.0.	372.0	281.2	309.5	285.3	247.4
	177.8	102.2	258.4	283.6	318.2	315.0	298.6	331.7	312.0	310.8	259.6	223.2

MULTAN SUNSHINE HOURS DATA

	JAN	FEB	MAR	APR	MAY	JUN	JUL	AUG	SEP	OCT	NOV	DEC
2000	197	242.4	270.2	285.3	263.1	234.6	237.9	306.6	275.3	292.8	229.5	222.3
2001	193.6	220.3	293.7	256.9	249.2	213.3	253.2	291.7	278.9	278.1	246.5	233.7
2002	221.7	200	270.1	253.6	280	261	239.2	265.8	255.8	275.8	206.1	219.7
2003	187.6	207.5	228.6	292.1	277.1	269	224	249.3	252	290.8	228.7	180.4
2004	143.1	207.7	287.4	227	252.3	240	234.1	222.4	263.3	236	228.5	205.2
2005	214	127.1	248.1	295.1	284.3	268.8	266.5	279.2	268.6	300.9	252.2	293.6

2006	190.9	161.3	259.6	278.3	249.4	265.8	234	244.2	270.9	259.7	198.6	214.4
2007	233.3	145.4	250.3	295	245.6	----	----	280.7	262.3	293.4	222.5	215.2
2008	190.6	221.2	282.2	273.9	246.9	246.9	289.4	227	268.7	253.4	251.7	180.7
2009	208.9	201.3	239.4	278.3	298.3	263.4	218.5	274.5	272.3	273	172.6	209.1

HYDERABAD SUNSHINE HOURS DATA

	JAN	FEB	MAR	APR	MAY	JUN	JUL	AUG	SEP	OCT	NOV	DEC
2000	263.0	274.9	282.6	219.6	309.8	179.7	197.7	233.6	248.4	298.7	275.4	278.8
2001	284.2	254.1	276.8	291.2	285.0	158.5	100.4	192.4	275.7	296.5	276.1	263.8
2002	271.2	259.5	273.7	288.1	303.3	250.3	187.5	191.3	282.2	276.0	260.8	254.8
2003	252.6	240.2	259.7	288.0	286.1	222.2	122.4	202.1	255.8	250.0	262.9	247.6
2004	263.0	275.1	313.2	265.9	273.6	216.4	199.3	189.1	271.1	250.4	260.6	244.9
2005	264.9	209.1	291.1	295.1	316.7	293.1	247.1	259.0	245.7	310.6	272.4	272.6
2006	262.8	251.8	289.1	291.2	308.1	258.3	144.8	180.3	271.5	292.5	265.0	242.3
2007	267.9	251.4	280.0	323.4	320.7	260.2	198.9	247.0	280.3	313.5	281.0	264.5
2008	230.1	265.5	291.7	286.9	293.3	184.1	151.6	242.8	249.6	293.5	277.9	212.2
2009	233.0	253.9	296.2	303.2	301.8	275.0	233.0	274.0	290.7	297.3	259.6	227.8

Table 8. Sun Shine Hours Data[3]

Author details

Ihsan Ullah[1], Mohammad G. Rasul[1], Ahmed Sohail[2], Majedul Islam[3] and Muhammad Ibrar[4]

1 Central Queensland University, School of Engineering and Technology, Rockhampton, Queensland, Australia

2 College of Electrical & Mechanical Engineering, Department of Mechanical Engineering, Rawalpindi, Pakistan

3 Queensland University of Technology, Brisbane, Australia

4 Islamia College University, Department of Physics, Peshawar, Pakistan

References

[1] Water & Power Development Authority of Pakistanhttp://www.wapda.gov.pk/htmls/power-index.html,June 09, (2012).

[2] Alternative Energy Board (AEDB)Power sector situation in Pakistan, Alternative Energy Board (AEDB), Deutsche Gesellscharft fur Technische Zusammenarbeit (GTZ) GmbH, September (2005).

[3] Pakistan Meteorological DepartmentGovernment of Pakistan, May (2010).

[4] Raja, I. A, & Twidell, J. W. Statistical analysis of measured global insolation data for Pakistan, Renewable energy: (1994). , 4(2), 199-216.

[5] World Energy CouncilRenewable energy in South Asia: Status and prospects. London, UK; November (2000).

[6] Private Power Infrastructure BoardMinistry of Water & Power Govt. of Pakistan

[7] Klaiss, H, Kohne, R, Nitsch, J, & Sprengel, U. Solar thermal power plants for solar countries- technology, economics and market potential. Appl Energy (1995). , 52, 165-83.

[8] Marketaki, K, & Gekas, V. Use of the thermodynamic cycle Stirling for electricity production. In: Proceedings of the 6th Panhellenic Symposium of Soft Energy Sources, (1999). , 283-290.

[9] Overview of solar thermal technologies by Centre for renewable energy resourcesUSA, May (1997).

[10] Teocharis TsoutsosVasilis Gekas, Katerina Marketabi, Technical and economical evaluation of solar thermal power generation, Renewable Energy 28 ((2003). , 873-886.

[11] Hargreaves, C. M. The Philips Stirling engine. Amsterdam: Elsevier Science, (1991).

[12] De Graaf, P. J. Multicylinder free-piston Stirling engine for application in Stirling-electric drive systems. In: Proceedings of the 26th Intersociety Energy Conversion Engineering Conference- IECEC '91. Boston, MA, USA, (1991). , 5, 205-210.

[13] Schlaich Bergermann und Partner GbREuroDish-Stirling System Description, Schlaich Bergermann und Partner GbR, Structural Consulting Engineers, Stuttgart June (2001).

[14] Li YaqiHe Yaling, and Wang Weiwei, Optimization of solar-powered Stirling heat engine with finite-time thermodynamics, Renewable energy 36 ((2011). , 421-427.

[15] Waqas NasrUllah Khan ShinwariFahd Ali, and A.H. Nayyar, Electric power generation from solar photovoltaic technology: Is it marketable in Pakistan? The Pakistan Development Review 43: 3(Autumn (2004). , 267-294.

[16] ClimateGlobal Warming, and Daylight Charts and Data; http://www.climate-charts.com/Locations/php,August (2010).

[17] National Renewable Energy Lab; http://wwwnrel.gov/analysis/tech_cap_factor.html, National Renewable Energy Lab USA, September (2010).

[18] Policy Framework and Package of Incentives for Private Sector Power Generation Projects in PakistanPrivate Power Cell, Ministry of Water and Power, Government of Pakistan, (1994). Annexure I, , 1.

[19] In the fieldEPRI Estimates Performance, Costs for Solar Thermal Plants, http://mydocs.epri.com/docs/CorporateDocuments/EPRI_Journal/Spring/1025049_InThe-Field.pdf,EPRI Journal Spring (2012).

[20] National Electric Power Regulatory AuthorityGovernment of Pakistan, http://nepra.org.pk/tariff_ipps.htm,November (2012).

Permissions

The contributors of this book come from diverse backgrounds, making this book a truly international effort. This book will bring forth new frontiers with its revolutionizing research information and detailed analysis of the nascent developments around the world.

We would like to thank Associate Professor Mohammad Rasul, for lending his expertise to make the book truly unique. He has played a crucial role in the development of this book. Without his invaluable contribution this book wouldn't have been possible. He has made vital efforts to compile up to date information on the varied aspects of this subject to make this book a valuable addition to the collection of many professionals and students.

This book was conceptualized with the vision of imparting up-to-date information and advanced data in this field. To ensure the same, a matchless editorial board was set up. Every individual on the board went through rigorous rounds of assessment to prove their worth. After which they invested a large part of their time researching and compiling the most relevant data for our readers. Conferences and sessions were held from time to time between the editorial board and the contributing authors to present the data in the most comprehensible form. The editorial team has worked tirelessly to provide valuable and valid information to help people across the globe.

Every chapter published in this book has been scrutinized by our experts. Their significance has been extensively debated. The topics covered herein carry significant findings which will fuel the growth of the discipline. They may even be implemented as practical applications or may be referred to as a beginning point for another development. Chapters in this book were first published by InTech; hereby published with permission under the Creative Commons Attribution License or equivalent.

The editorial board has been involved in producing this book since its inception. They have spent rigorous hours researching and exploring the diverse topics which have resulted in the successful publishing of this book. They have passed on their knowledge of decades through this book. To expedite this challenging task, the publisher supported the team at every step. A small team of assistant editors was also appointed to further simplify the editing procedure and attain best results for the readers.

Our editorial team has been hand-picked from every corner of the world. Their multi-ethnicity adds dynamic inputs to the discussions which result in innovative outcomes. These outcomes are then further discussed with the researchers and contributors who give their valuable feedback and opinion regarding the same. The feedback is then collaborated with the researches and they are edited in a comprehensive manner to aid the understanding of the subject.

Apart from the editorial board, the designing team has also invested a significant amount of their time in understanding the subject and creating the most relevant covers. They scrutinized every image to scout for the most suitable representation of the subject and create an appropriate cover for the book.

The publishing team has been involved in this book since its early stages. They were actively engaged in every process, be it collecting the data, connecting with the contributors or procuring relevant information. The team has been an ardent support to the editorial, designing and production team. Their endless efforts to recruit the best for this project, has resulted in the accomplishment of this book. They are a veteran in the field of academics and their pool of knowledge is as vast as their experience in printing. Their expertise and guidance has proved useful at every step. Their uncompromising quality standards have made this book an exceptional effort. Their encouragement from time to time has been an inspiration for everyone.

The publisher and the editorial board hope that this book will prove to be a valuable piece of knowledge for researchers, students, practitioners and scholars across the globe.

List of Contributors

R. Mahamud, M.M.K. Khan, M.G. Rasul and M.G. Leinster
Central Queensland University, School of Engineering and Built Environment, Rockhampton, Queensland, Australia

Audai Hussein Al-Abbas
Foundation of Technical Education, Al-Musaib Technical College, Babylon, Iraq

Jamal Naser
Faculty of engineering and Industrial Sciences, Swinburne University of Technology, Hawthorn, Victoria, Australia

M.N. Lakhoua
Université de Tunis El Manar, Ecole Nationale d'Ingénieurs de Tunis, LR11ES20 Analyse, Conception et Commande des Systèmes, Tunis, Tunisie

M. Harrabi and M. Lakhoua
Société Tunisienne de l'Electricité et du Gaz, Tunisie

A. Yu. Ryabchikov
"Turbines and Engines" department, Urals Federal University (UrFU), Yekaterinburg, Russia

Adnan Moradian
Ministry of Power, Niroo Research Institute, Tehran, Iran

Farid Delijani and Fateme Ekhtiary Koshky
Ministry of Power, East Azarbayjan Power Generation Management Co., Thermal Power Plant of Tabriz, Tabriz, Iran

Gurdeep Singh
Department of Environmental Sciences and Engineering, Indian School of Mines, India

A.K.M. Sadrul Islam
Department of Mechanical and Chemical Engineering, Islamic University of Technology (IUT), Gazipur, Bangladesh

Md. Ahiduzzaman
Department of Agro-processing, Bangabandhu Sheikh Mujibur Rahman Agricultural University, Gazipur, Bangladesh

Ihsan Ullah and Mohammad G. Rasul
Central Queensland University, School of Engineering and Technology, Rockhampton, Queensland, Australia

Ahmed Sohail
College of Electrical & Mechanical Engineering, Department of Mechanical Engineering, Rawalpindi, Pakistan

Majedul Islam
Queensland University of Technology, Brisbane, Australia

Muhammad Ibrar
Islamia College University, Department of Physics, Peshawar, Pakistan

Printed in the USA
CPSIA information can be obtained
at www.ICGtesting.com
JSHW011352221024
72173JS00003B/263